Encoding Bioethics

Encoding Bioethics

AI in Clinical Decision-Making

Charles Binkley
and
Tyler Loftus

UNIVERSITY OF CALIFORNIA PRESS

University of California Press
Oakland, California

Cataloging-in-Publication data is on file at the Library of Congress.

ISBN 978-0-520-39752-1 (cloth : alk. paper)
ISBN 978-0-520-39753-8 (pbk. : alk. paper)
ISBN 978-0-520-39754-5 (ebook)

Manufactured in the United States of America

33 32 31 30 29 28 27 26 25 24
10 9 8 7 6 5 4 3 2 1

To David, Jake, and Grace—
My Way, My Truth, My Life (CB)

To my family,
whose love and support are endless (TL)

CONTENTS

ILLUSTRATIONS

PREFACE

Our story begins where all good health-care stories begin: in the trenches of patient care. We—the authors—have many professional roles in health care, but above all, we are surgeons. Providing the best patient care possible inside and outside the operating room is our true north and deep passion. We have taken different paths toward this shared goal, Charles by caring for patients suffering from cancer while developing expertise in clinical, organizational, and technology ethics, and Tyler by caring for patients suffering from traumatic injury and other emergency surgical conditions while developing expertise in artificial intelligence (AI)-enabled clinical decision support. Throughout our careers, it has become increasingly apparent that providing the best patient care possible has many facets and takes a village.

Health care, like many other industries, is increasingly subspecialized and siloed. Many cathedrals—renowned for their architectural and engineering excellence—were built over the lifetimes of one or two master builders who intimately understood and personally oversaw every critical aspect of the project.

The results were breathtakingly beautiful, but the process was slow and highly dependent on a few people overseeing many workers who were granted little agency. Today, great architectural feats are accomplished by large teams comprising subspecialists in every aspect of architecture and engineering, working in concert to deliver excellent results efficiently. Modern health care intends similar agility and cooperation among subspecialists, but instead we observe that it is often fragmented and occasionally fractured.

Herein, we endeavor to unite perspectives from major stakeholders of ethical AI in clinical decision-making: patients, physicians, developers, payers, and health system administrators. This unification is not a roadmap for training AI ethics master builders, but rather to generate a shared understanding of the values and principles of each stakeholder. Through this understanding, we hope that each stakeholder will make decisions and take actions that build stronger, more ethical applications of AI-enabled clinical decision support, ultimately toward the best patient care possible. We envision a brighter future in which human-AI teams address ethical challenges that we face in our practices: a primary care physician is notified by a computable phenotyping algorithm that conditions for implementing a living will are imminent, prompting documentation of the living will and ensuring that the patient's wishes are honored; a multidisciplinary tumor board consults with AI-enabled decision support in recommending the optimal sequence and timing of chemotherapy, radiation, and surgery for a patient suffering from pancreatic cancer; a blood bank manager collaborates with a resource allocation algorithm to divert blood products away from a futile massive transfusion and toward a patient who is likely to benefit from transfusion. Fulfillment of this vision will require that siloed efforts coalesce as stakeholders understand one another and their own place in a reimagined and stronger health-care system.

Besides caring for patients, our other passion is recruiting, educating, and mentoring future clinicians and nonclinician health-care leaders. It is insufficient that we only teach clinical skills to future clinicians. With the rapid introduction of technological advances, such as AI, at the bedside and into the operating room and the clinic, it is essential that future clinicians know how to maximize the benefit and minimize the harm to patients from these tools just as much as from the scalpel. In addition to clinicians, developers, programmers, data scientists, and engineers are becoming even more essential members of the health-care team. Embracing these new members as partners in patient care requires that they share in common with their clinician colleagues certain ethical commitments. First among these is the provision of high-quality, patient-centered care. Creating a common understanding of ethical values between clinicians and nonclinicians in caring for patients is one of our primary motivations for writing this book.

Because we each play roles outside of direct patient care, Charles as an ethicist and Tyler as an AI developer, we wrote this book not only for those who are, or will be, actively engaged in caring for patients, but also for those who seek to deepen their knowledge about the use of AI for clinical decision-making. It is our intention that students of any discipline, particularly philosophy, ethics, engineering, and technology, will find our applied approach to often lofty and complex concepts accessible. In particular, we wrote the cases at the end of each chapter with an eye toward diverse audiences who can use the accompanying questions to spur discussion and the exchange of ideas.

We also realize that there is a significant risk that AI will make medicine and health care even more impersonal and seemingly transactional. It is often up to the health-care leaders, executives, and payers to decide to what extent heath care is accessible, equitable, and human, and to what extent it is depersonalized, cost-centered, and mechanistic. We believe that the ethical application of AI in health care presents an opportunity for the future to be both human and efficient, caring and cost-effective, personal, and value based. We hope that what we have written will help those leaders align their very important ethical duty to prepare and plan with their equally important duty to truly care about each and every patient admitted to their hospital, treated in their clinic, or enrolled in their health plan. Perhaps it will be through the use of ethically developed and programmed AI for clinical decisions that the values and preferences of patients and their physicians will finally find alignment with the priorities of health systems and payers.

ACKNOWLEDGMENTS

Just as it takes a village to take care of a sick patient, so too does it take a village to write a book about bioethical considerations in AI.

Charles would like to acknowledge his colleagues at Hackensack Meridian Health and Hackensack Meridian School of Medicine, especially Dr. Hannah Lipman, Laurel Hyle, Dr. Miriam Hoffman, Dr. Bryan Pilkington, Brenda Rivera, Dr. Lauren Koniaris, Dr. Bommae Kim, and the Medical Library Staff; his colleagues at the Markkula Center for Applied Ethics, especially Dr. Brian Green, Irina Raicu, Ann Skeet, and Dr. Don Heider; Dr. Mark Sendak from the Duke Institute for Health Innovation; his coauthor Dr. Tyler Loftus for embracing this project with passion and insight; and Chad Attenborough and Dr. Chloe Layman from the University of California Press, who have been gentle, wise, kind, and enormously appreciated at every step along the way.

Tyler would like to acknowledge his colleagues, mentors, and sponsors at the University of Florida, especially Dr. Azra Bihorac, Dr. Parisa Rashidi, Dr. Gilbert Upchurch, Jr., Dr. George

Sarosi, Jr., Dr. Philip Efron, Dr. Raymond Moseley, and Dr. William Hogan; Dr. Randall Moorman from the University of Virginia; Dr. Christopher Tignanelli from the University of Minnesota; the world-class surgical residents and fellows at the University of Florida, whose passion and commitment for providing the best surgical care possible is a limitless source of strength and inspiration; Dr. Charles Binkley for sharing his deep commitment to bioethical applications in the trenches of patient care; and Chad Attenborough and Dr. Chloe Layman from the University of California Press for sharing their wisdom and advice in shepherding this project toward a fulfilling conclusion.

CHAPTER ONE

Introduction to AI Clinical Decision Support Systems

Ethical Considerations

Human beings are somewhat unique in their caregiving for their own sick and diseased.[1] Around this seemingly preternatural urge has developed a tremendously sophisticated medical system wherein the sick are diagnosed and treated. It has long been acknowledged that the tools wielded by medical practitioners have the potential for great benefit and also for great harm, depending on how they are used. The scalpel, for instance, can be used to remove cancerous tumors to extend life. The same scalpel can cause irreparable harm and death. While every tool used in medicine has the potential to be directed either toward healing or harming, it is through a professed system of ethics that patients can trust that practitioners of medicine aim their tools toward the benefit, and away from the harm, of their patients.[2]

MAXIMIZING BENEFIT AND MINIMIZING HARM: THE ROLE OF ETHICS

It seems that the same thing can be said of artificial intelligence (AI) systems: they are capable of great benefit and also great harm. In health care generally, AI systems may be beneficial by increasing access to care, making care more affordable, improving the overall quality of care, and fostering increased trust in the health-care system. On the flip side, the systems may also further entrench a two-tiered health-care system in which the insured get high-quality, personalized care and the uninsured and underinsured, who are often patients of color, get lower quality and often depersonalized care.[3]

AI systems also have the potential to dehumanize medical care and devalue the role of the therapeutic physician-patient relationship. AI systems that are able to independently make a diagnosis and prescribe a treatment are likely to increase the efficiency and decrease the cost of providing health care. Such systems could plausibly replace many of the roles played by physicians. It is likely that some of the functions performed by physicians today will be objectively improved when physicians collaborate with AI systems, and some functions may actually be performed better by AI systems alone. However, at its core, medicine is relational, and most patients place a high value on the relationship. Some patients may even value the relationship so much that they would accept a less perfect human to a more perfect AI system. Although AI systems may objectively perform some functions better than physicians, the relationship between the patient and their physician cannot be replaced by an AI system, regardless of how humanoid the system is. Furthermore, if patients perceive that increased efficiency and decreased costs have led to the elimination of the physician-patient relationship, they may come to further distrust health-care institutions.

As AI systems disseminate into almost every aspect of society, particularly health care, one must ask how to ensure that their potential for good is being maximally actualized and the harm that they could cause is being assiduously avoided. Multiple frameworks and checklists have been developed and promulgated that seek to provide a structure for these systems that, if observed, would shift the balance toward benefit and away from harm.[4] One of the issues becoming apparent in the process of adopting a framework for AI ethics is that these systems, in and of themselves, have no ethical obligation other than to perform as they claim. They are essentially amoral systems without a predetermined set of obligations or duties.

At this point, comparisons between AI and medicine diverge, since medicine has a long, well-established, and in many cases legally codified system of ethical obligations and duties. Before embarking on the application of these duties, it is helpful to first articulate what these ethical obligations are and how they are understood by most physicians today. Perhaps the best known and most widely utilized medical ethics approach is that described by Tom L. Beauchamp and James F. Childress in *Principles of Biomedical Ethics*.[5] The system these authors describe has come to be known as *principlism* and encompasses four principles: respect for patient autonomy, beneficence, nonmaleficence, and justice. Applied to medical decision-making, these four principles guide physicians both in their own medical decision-making and also when engaging with the patient in shared decision-making.

USING ARTIFICIAL INTELLIGENCE FOR MEDICAL DECISION-MAKING

It is within this ethico-medical context that AI systems are being introduced. Although AI systems are being implemented in almost every facet of health care, including operational and administrative arenas, this text specifically addresses the use of AI systems that are used to make clinical decisions and that involve both physicians and patients.[6] These systems, often called artificial intelligence clinical decision support (AI CDS), are trained on historic patient information to make a desired clinical prediction. For instance, in designing a system that will predict which patients are at increased lifetime risk of developing kidney disease. The system will (ideally) have access to hundreds of thousands of patient records and either make a prediction based on outcome labels determined by human users (supervised learning) or discover patterns associated with its own computer-determined outcome labels (unsupervised). These systems learn from the medical histories and outcomes of past patients to predict clinical outcomes, disease classifications, and therapeutic responses for current and future patients. Importantly, and a topic of discussion later in the text, current patients can also contribute their health information to the system after it has made their prediction. In doing so, the system continues to learn new associations, which can shift over time (e.g., the COVID-19 pandemic affected time-honored associations between fever and bacterial sepsis).

This work focuses on AI CDS specifically for three important reasons.[7] First, a general AI ethics framework that seeks to address every AI application would inevitably be either so general as to lack necessary granularity for certain applications or

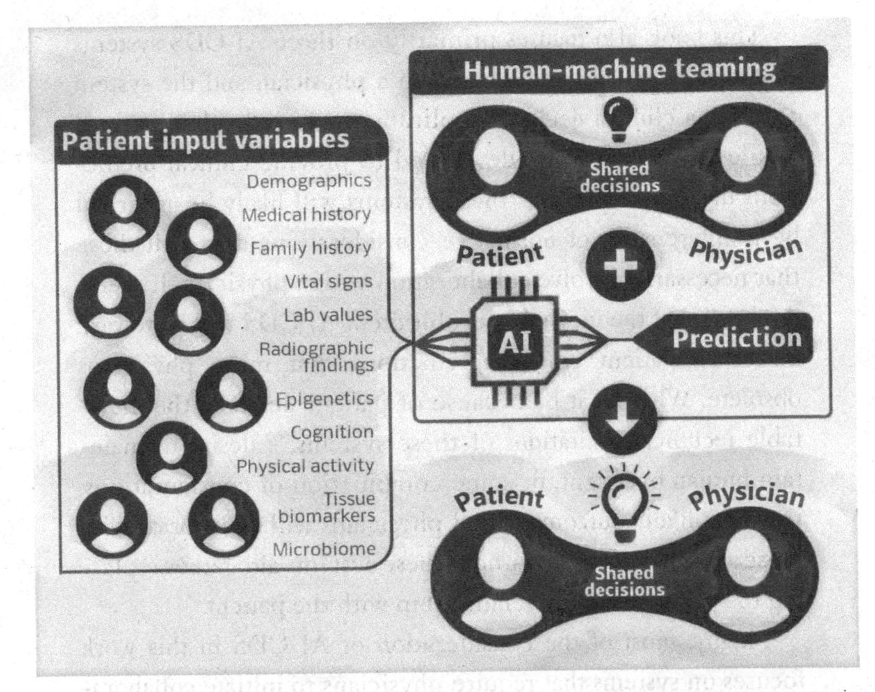

Figure 1. How AI CDS works

so prescribed and detailed as to be overly restrictive to other applications. Second, medicine generally, and physicians specifically, already have well-defined and widely understood ethical codes. There is no need to reinvent or rediscover medical ethics; rather, medical ethics must be applied to AI systems that are used for medical decision-making. Third, these systems are being introduced into a human relationship between a physician and a patient and are best understood in that context. The relationship is built on the patient's trust in their physician. The patient's trust in the physician is justified and perpetuated because of the physician's profession of ethical obligations to the patient.

This book also focuses primarily on those AI CDS systems that require collaboration between a physician and the system to make a clinical decision applicable to an individual patient. While AI CDS is being developed to provide clinical predictions directly to patients, these systems will likely be governed by a different set of normative considerations than will those that necessarily involve collaboration with a physician. It seems doubtful that the inevitable evolution of AI CDS will be toward direct to patient clinical predictions that make physicians obsolete. Whether it be because of patient demand, the inevitable technical limitations of these systems, a desire to maintain human oversight, or some combination of considerations, the most likely outcome is that physicians will collaborate with these systems routinely without these systems altogether replacing the physician in the relationship with the patient.[8]

Finally, most of the consideration of AI CDS in this work focuses on systems that require physicians to initiate collaboration with the AI system and volitionally request clinical decision support. The AI decision is offered to physicians "on demand" when physicians request it. This is in contrast to AI CDS systems that are "pushed out" to physicians without the physician requesting collaboration. The ethical relevance of this distinction has several considerations. First, on demand clinical decisions presume that the physician is initiating the collaboration with a specified clinical aim, such as making a diagnosis, prescribing a treatment, or predicting an outcome. Second, the physician is prepared to act on the prediction, typically by accepting or rejecting it. Last, the physician is prepared to inform the patient of the prediction made by the system in the process of shared decision-making, even if the physician does not inform the patient that the decision was made in collaboration with an

AI system. In contrast, when physicians receive a clinical prediction without requesting it, they may be unprepared to assess and, in turn, act on the information and inform the patient. This could lead to unintended use of the prediction; failure to act on the prediction, even if it would be clinically beneficial to the patient; and failure to inform the patient of the prediction.

THE STAKEHOLDERS IN THE PHYSICIAN-PATIENT-AI RELATIONSHIP

Just as each clinical ethics dilemma involves the perspective and values of many different stakeholders—the patient, the physician and medical team, the patient's family, the hospital and health system in which the patient is receiving care, the community to which the patient belongs, the payer of the patient's hospital bill, and others—so too are their multiple stakeholders involved in the development, programming, deployment, use, and auditing of AI CDS. Although each of these AI CDS stakeholders has a different perspective and also a different value system, they all have a common relationship to the AI CDS: entering into the physician-patient relationship.

How, then, can it be assured that the AI CDS is directed toward the patient's benefit and away from harm? It is through the expressed ethical commitment of each of these stakeholders, whom we define as the patient, the physician, the programmer, and the health system administrator and payer, that patients can be assured that AI CDS maximizes benefit and minimizes harm.

This book seeks to consider the pertinent ethical concerns, derived from the physician-patient relationship, that are relevant to each stakeholder. Beauchamp and Childress's four principles are a common starting point for understanding the ethical

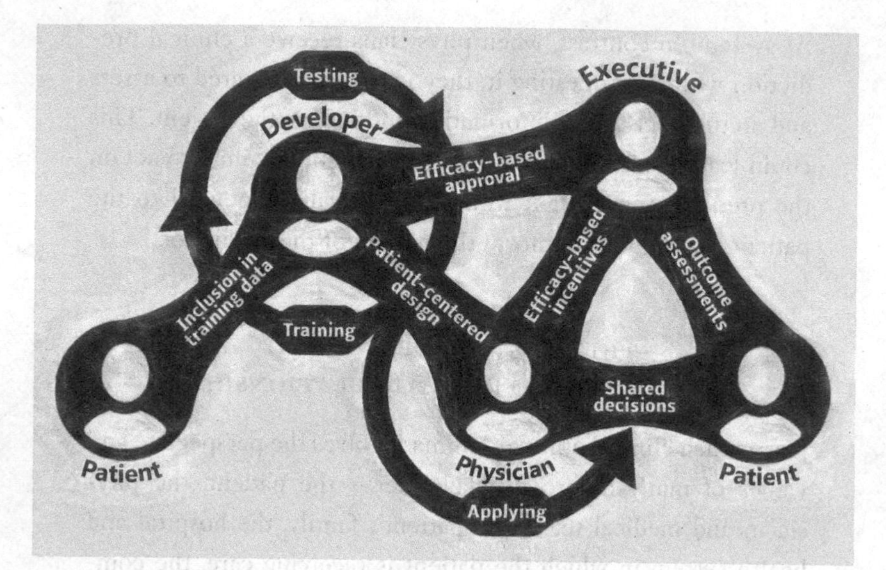

Figure 2. Stakeholders in AI CDS

concerns relevant to each stakeholder. This is not because principlism is particularly well suited to AI CDS, but rather because it is the most commonly used and best-known framework in medicine today.[9] In fact, as this text highlights, the four principles have several weaknesses generally and specifically when applied to AI CDS.[10] For instance, the principle of autonomy is often thought of as encompassing informed consent, veracity, and confidentiality. While it is true that these other principles may demonstrate respect for patient autonomy, when it comes to the ethical issues facing stakeholders involved with AI CDS, informed consent, veracity, and confidentiality are important considerations in addition to, and distinct from, respect for autonomy.

When considering the stakeholders who have a relationship with AI CDS, it is important to acknowledge that the physician has a direct relationship with both the patient and AI CDS, whereas the other stakeholders have direct relationships with AI CDS but indirect relationships with patients. The importance is twofold. First, it is into the physician-patient relationship that both AI CDS and the other stakeholders are introduced, a relationship governed by ethical considerations. By entering into a normative relationship, these other stakeholders are reasonably expected to not only understand the ethical norms of the relationship but also abide by them. Second, it is the physician whom the patient knows and sees, and with whom the patient talks and has the primary relationship. This is true even though other stakeholders may play an outsized role in the medical decisions made within the physician-patient relationship. The physician is vulnerable not only to the patient's ire, but also to peer review, licensing boards, tort claims, and professional societies if the patient is harmed.

Take, for instance, the decision by health systems to sell patient health information to AI developers. A patient discovers that their health information has been sold and used to profile them across AI platforms—not by name or social security number, or medical record number, but by their own unique combination of health information that is specific to them and distinct from everyone else. The patient may be furious with their physician, blaming them for dereliction of their ethical obligation to maintain confidentiality, resulting in loss of trust. The physician, however, may be unaware that the health system sold patient data. They may have no idea how to respond to the patient. In fact, even the physician's health data may have been sold. Patients may feel similarly angry and betrayed by their physician if they learn that AI CDS has been used to make high-stakes clinical predictions about them without their being informed or giving consent. Although the physician has the most "skin in the game," when nonphysician stakeholders enter into the relationship between the physician and the patient, the other stakeholders should abide by the ethical norms that govern the patient-physician relationship.

NORMATIVE CONSIDERATIONS IN HUMAN RELATIONSHIPS AND AI COLLABORATIONS

While myriad ethical frameworks are proposed for AI, this book takes a relational approach since medicine is ultimately relational. Applying a relational way of thinking about AI ethics could serve as a model for normative considerations of other AI applications. Humanity is ultimately relational, and patient-physician relationships are merely one kind of relationship, which occurs between one human experiencing illness and another human who has the knowledge and skill to treat the illness or promote greater health. While the normative considerations in medicine are well defined, in other human relationships the obligations are less well defined, but just as important. Introducing AI systems into society will require that humans understand not only collaborations between AI systems and humans, but also the relationships between humans to which the AI-human collaboration is most similar.

Many of the characteristics of the physician-patient relationship described in this book are presented as they would ideally be practiced. However, as health care has become increasingly commercialized it has, in many instances, become less personal and more transactional. One risk when introducing AI CDS into health systems is that systems that maximize cost savings, increase efficiency, and improve reimbursement will be prioritized for development and deployment over systems that are purely clinically beneficial. It may also be that decisions about which AI system to develop or deploy are not made with input from the physicians collaborating with the system, nor from the patients and families affected by the system, but rather by executives who are responsible for the hospital or health system's profitability. That is not to say that financial security is not a legitimate consideration and even ethically relevant. Rather, the concern is that the potential of AI CDS to improve clinical decision-making and overall patient care will be thwarted by the potential to contain costs and improve profit margins.

If concerns about AI CDS being used primarily to drive profit can be adequately addressed, then perhaps these systems could lead to the reinvigoration of the art of medicine. While it may seem paradoxical that an advanced technology would enhance the humanistic aspects of medicine, there are two important ways in which it has the potential to do so. First, as others have highlighted, AI systems can reduce the rote nonclinical activities that consume physicians' time and energy. These activities are more burdensome and less professionally rewarding than caring for patients, which is a physician's highest calling. However, there is another important role that AI systems have the potential to play that will increase the human dimensions of the physician-patient relationship. The corollary of this is that these systems have the potential to significantly reduce, but not altogether eliminate, clinical uncertainty. This will allow physicians to engage with patients and families in shared decision-making and to focus less on complex calculations and predictions based on error-prone hypothetical-deductive reasoning. Physicians will be able to focus more on problem solving, creativity, altruism, and the art of medicine, which is extremely important from a relational perspective. As medicine has evolved, the middle area between life and death has greatly expanded. This middle area has led to tremendous clinical uncertainty.

THE EMOTIONAL AND MORAL BENEFIT OF REDUCING CLINICAL UNCERTAINTY

Patients are often offered high-risk treatments because the alternative, doing nothing, would certainly result in death. However, "doing something" has too often also resulted in death, and post-treatment death usually involves pain, suffering, hospitalization, and exorbitant medical bills. Physicians offer high-risk treatments because they are uncertain whether the patient would benefit, and because they do not have the information and tools to make an accurate and personalized prediction of the potential risk and benefit of the treatment for an individual patient. For instance, patients have also been maintained in an unconscious state for years, sometimes decades, not because they and their families value life in any form, nor because they expected a miracle, but rather because physicians have not been able to accurately predict a patient's potential for recovery after a severe neurologic injury. Conversely, patients have also been refused potentially life-prolonging treatments because, more often than not, a treatment was unsuccessful, and physicians could not justify offering the treatment two hundred times unnecessarily for one patient's benefit. Again, physicians have not been able to predict with precision who would benefit and who would not benefit from a given treatment.

The increasing clinical uncertainty that has accompanied medical complexity has shifted the role of the physician from active participant in shared decision-making to that of soothsayer and oracle. Divinely inspired, these roles may seem powerful and grandiose, but for mere humans, they can be morally burdensome and deeply anxiety provoking. In an attempt to improve on their own epistemic authority, physicians have turned to scoring systems and clinical predictors to augment clinical reasoning and to help them decide for which patients the benefit would outweigh the risk and for which patients the risk would outweigh the benefit. Not only have these tools provided a defense for the physician to the patient and their family, their peers, and the courts in case harm has come to the patient, but they also may have assuaged the conscience of physicians seeking to reassure themselves that they haven't unnecessarily harmed another human contrary to the oath they have taken.

Patients and their families are asked to make extremely high-stakes decisions based on the professional judgment of one physician, or in some cases, such as tumor boards and case conferences, a small group of physicians. The ambiguity of not knowing, and not being able to know, in which direction your own or your loved one's illness trajectory is headed, either toward recovery or toward death, and over what span of time, has been the source of great distress for patients and their families. This distress has led to conflicts within families and between patients and their families and physicians. Although the patient and family suffer most from clinical uncertainty, ambiguity, and trials of treatments, physicians, who value harmonious relationships with their patients and their patients' families, also suffer.

When a seriously ill patient exists in the gray zone between life and death, and it is unclear whether the patient will recover and thrive, remain alive but unconscious, or die, physicians will often offer a trial of treatment to better predict the direction in which the trajectory of the patient's life is going. These periods of uncertainty are not only distressing for patients, families, and physicians, but they are also costly for the health-care system. An often-cited statistic is that somewhere between 10 and 25 percent of health-care expenses occur in the last year of life.[11] Inevitably, clinical uncertainty and trials of treatment account for much of this expense, the cost for which is shared by other patients and society.

REALIZING THE PROMISE OF AI FOR CLINICAL DECISION-MAKING

With the advent of AI-enabled clinical predictions there is the potential to significantly reduce clinical uncertainty across the care continuum. For patients and families, this could mean less distress and more objective information on which to make high-stakes decisions.[12] For health systems and payers, this could mean significant savings that could be directed toward providing medical care for more people. For physicians, this could mean shifting their function from prognostication calculator to wise counselor and partner in decision-making, thus unburdening them of a role that has become increasingly perilous and detracts from the deeper, more meaningful aspects of caring for patients.[13]

This book proposes a framework for introducing AI into clinical decision-making that realizes the potential of AI systems to do good and avoid harm while attending to ethical concerns. This framework depends on all AI CDS stakeholders understanding and respecting the centrality of the physician-patient relationship and the ethical norms that protect it. While rules, regulations, and laws are one means of assuring that stakeholders in AI CDS abide by ethical standards, another is by acknowledging the importance and uniqueness of the relationship into which the stakeholders are entering, with the patient at the center, a role that one day each of the stakeholders will also occupy.

CHAPTER TWO

The Physician and AI Clinical Decision Support Systems

Introducing AI CDS into the physician-patient relationship raises ethical concerns applicable to physicians. Physicians are ethically bound, first and foremost, to do good and avoid harm to their patients. In addition, physicians must show respect for their patients as autonomous persons, maintain patient privacy, be truthful to their patients, and engage with their patients in shared decision-making and informed consent. Ethical concerns arise when AI CDS shares with physicians in decision-making for patients, particularly when the decisions made by AI CDS are unexplainable and the reasoning behind the decision cannot be understood by physicians.

This chapter systematically compares the ways in which physicians currently make decisions—by referring to a textbook or journal article, using a predictive clinical calculator, and consulting with a colleague—to the use of AI CDS for shared decision-making. The goal of such a comparison is to highlight the distinct ethical issues raised by CDS and also to better define the nature of the collaboration between physicians and AI systems. The chapter concludes by offering suggestions for how the concerns might be addressed before the systems are widely deployed.

THE PHYSICIAN-PATIENT RELATIONSHIP

Several features of the physician-patient relationship make it unique and are important for the ethical considerations of using AI or machine learning (ML) in the clinical context. Perhaps the most striking is the vulnerability of the patient in the relationship. Multiple factors contribute to the patient's position. For one, the patient is often sick and suffering when they enter into the relationship with a physician. There is also typically a significant gap in medical knowledge between the patient and the physician. The physician often holds the means by which the patient will be healed and their suffering ended. The physician also enters into the relationship in a defined and professional role, not primarily as a fellow human being, but as one with epistemic authority.

The Ethical Obligations of the Physician-Patient Relationship

In order to safeguard the relationship between the physician and the patient, there are specific obligations owed to the patient by the physician. Chief among these duties is that the physician must do good and avoid harm to the patient.[1] Physicians do this by directing their clinical decisions toward the good of the patient. Recognizing that the powerful tools of the physician could be used for either good or harm, this duty protects the vulnerable patient by establishing the duties of the physician in their relationship with the patient.[2]

In addition, physicians must show respect for their patients as autonomous persons. At the most basic level, respect for patients as individuals protects them against a paternalistic model in which the physician knows best and the patient must do what they are told without explanation or understanding. Physicians engage in a model of shared decision-making with patients wherein the physician either makes a recommendation or presents a range of possible options. The physician's duty is to explain the options, their pros and cons, and help the patient come to a decision by answering questions and discussing concerns.[3]

Physicians further show respect for their patients by being truthful and transparent with them. This commitment to honesty goes beyond the minimal requirement that physicians not lie to patients. Physicians must disclose a patient all information that is relevant to their care, particularly information that may influence the patient's decision-making.

Finally, the physician should not have a competing primary interest besides the patient's benefit, and if there are competing interests, at the very minimum the patient should be informed. For instance, the patient should be informed if the physician who is treating them with a medication for cancer is also a major shareholder in the company that makes the medication.

A physician's public profession to do good, avoid harm, and respect their patients as individuals is the foundation of the doctor-patient relationship and provides the basis for patients to trust their physicians.[4] This covenant, of sorts, is assured not only by the physician's own personal integrity but also through professional societies and government oversight.

INTRODUCING AI/ML DECISIONS INTO THE PHYSICIAN-PATIENT RELATIONSHIP

Against the backdrop of the physician's relationship with the patient and the obligations that derive from that relationship, one can better understand how to consider what AI/ML systems mean to the physician and to their relationship to the patient. The first point of consideration is that using AI/ML to support clinical decisions introduces another entity into the physician's relationship with their patient. In many cases, the AI/ML system will be opaque to the patient and will function largely "behind the scenes" to support either a unilateral decision that the physician will make or a recommendation that the physician will make to the patient.

For instance, intraoperative AI/ML systems that predict the optimal placement of a prosthetic knee provide support for technical decisions that are made by surgeons. The patient is not typically involved in deciding the proper alignment of their prosthetic joint. This decision is left to the surgeon. Under usual circumstances, the surgeon would only be asked to give an account of their decision-making process if the patient suffered a complication, for example, if the prosthetic joint became detached from the surrounding bone or the patient were left with a severe limp after surgery. In these cases the patient, or a peer if the case were referred for quality review, or an attorney if the case were litigated, might ask the surgeon to explain their reasoning for aligning the joint as they did.

In other instances, AI/ML systems will support clinical decisions that the physician will then recommend to the patient. For instance, a system might recommend a specific chemotherapy regimen for treating breast cancer, including which drugs and what dose. The physician will then make the treatment recommendation to the patient.[5]

These examples illustrate one of the key ethical concerns that physicians will have when AI/ML systems are introduced into clinical decision-making, namely, that most AI/ML decisions are not understandable by the physician because the reasoning underlying the decision is not known, the so-called black box problem.[6] Because physicians will not be able to comprehend how the system made a specific clinical recommendation for that specific patient, physicians will in turn not be able to judge whether either the clinical decision they are making unilaterally or what they are recommending to the patient in order to engage in shared decision-making fulfills their obligation to do good and avoid harm.[7]

Understanding Physician-AI Collaboration for Clinical Decision-Making

When an AI/ML system makes a clinical decision and in turn communicates that decision to a physician, regardless of whether the decision is one the physician will make unilaterally or present to the patient to engage in shared decision-making, the system and the physician are collaborating in order to make a clinical decision for a patient. Defining the nature of that relationship is important for establishing that the ethical obligations owed to the patient by the physician apply to the system also. One way of both better understanding that relationship and at the same time examining the ethical concerns that this new relationship between physician and AI system raises is to examine the process of physician decision-making and compare that process to the use of AI systems for clinical decision-making.

The AI system, based on its ex ante programming, training, expert annotation, and possibly ongoing learning, will use information in the patient's electronic medical record (EMR) to make an intentional prediction in the form of a clinical decision for the patient.[8] The physician, typically simultaneously, will gather data from the patient and their EMR and, based on the physician's training, education, and experience, also make a clinical decision. The physician may also refer to a journal article or textbook, use a clinical calculator, or confer with a colleague in arriving at their decision. After the system communicates its decision to the physician, the physician must then decide what action to take. It is at this point that ethical concerns arise based on the interaction between the system and the physician in making a clinical decision on behalf of a patient.

Mapping Ethical Concerns about Physician-AI Collaboration for Making Clinical Decisions

Comparing the interaction between physicians and AI systems to the three typical means that physicians employ to make medical decisions will highlight the distinct ethical concerns that arise from using AI systems that support clinical decisions. Physicians may turn to some source of medical information, be it a textbook, journal article, or website, in order to assist them in making a clinical decision. The information is typically based on clinical or basic science research, or best practice. It is often written by other physicians, or groups of medical experts, who have the same general professional responsibility as the physician who is reading their material.

There are also clinical calculators available to help guide physicians in their decision-making. A common example is a surgical risk assessment tool, into which pertinent patient parameters, such as age, comorbidities, and labs are entered, and which then predicts the patient's risk of death or serious complication if they undergo an operation.[9] These calculators make a prediction based on pooled data from patients who have previously undergone surgery and had their outcome in terms of survival and complications reported. The risk stratification tools have usually been verified to test their predictive accuracy.

Finally, physicians turn to their colleagues for help in making clinical decisions. These can be formal processes, such as a difficult case conference or a tumor board, or informal, such as when a physician runs a case by a colleague, often called a "curbside" consult. Pertinent patient data are shared, and the patient is discussed among colleagues until a conclusion or decision is reached.

Looking carefully at how these familiar models of collaborative decision-making differ from the use of AI/ML clinical support systems will highlight some of the most important ethical issues physicians will face as these systems are introduced into their practice. It will also allow the relationship between the AI system and the physician to be better understood.

RESOURCES FOR PHYSICIAN DECISION-MAKING: MEDICAL INFORMATION

Using AI/ML systems differs from consulting textbooks and journal articles to make clinical decisions in several important ways. At the most basic level, textbooks and journal articles provide physicians with information on which physicians can make a judgment, whereas AI systems used to support clinical decisions provide a judgment about what is best for the patient without using supporting information.[10] Moreover, the judgments of the systems can continuously change as the system learns from the outcomes of its previous decisions. The physician must therefore be able to first trust that the system and physician together will do more good and avoid more harm than the physician would do alone, and that as the system evolves and learns from the results of it's predictions, it will continue to do more good and cause less harm than its previous iteration.[11]

Trusting AI Clinical Decisions to Do Good and Avoid Harm

Physicians can trust that the diagnostic and treatment advances published in the medical literature do good, avoid harm, and in sum, do more good and/or avoid more harm than whatever previous means of diagnosis or treatment they are replacing. The reason physicians can trust the advances and discoveries published in the medical literature is that they are the result of a clinical trial or other clinical research. The information has been evaluated, peer reviewed, and accepted as valid.

Physicians currently rely on two important sources of information to make clinical decisions: evidence-based medicine (EBM) and real-world data. EBM begins with a clinical question, such as whether chemotherapy agent A or B is more effective at extending disease-specific survival for patients with early stage breast cancer. Patients would typically be randomized to receive treatment with either chemotherapy agent A or agent B. As much as possible, the personal characteristics, such as age, comorbidities, and cancer stage, would be the same in patients assigned to receive agent A and agent B. The patients receive the assigned treatments and are followed to determine which treatment is superior at achieving the determined outcome. Real-world data rely on large databases, such as from insurance or Medicare claims, patient-reported outcomes, and patients' EMRs. A similar clinical question would be asked, such as whether chemotherapy agent A or B was more effective at extending disease-specific survival for patients with early breast cancer. The available data would be reviewed and outcomes would be compared in order to reach a conclusion.

It is important to note that medical advances are put into practice not just because they work as they claim to, but more importantly, because they are beneficial and not unacceptably harmful to patients. Moreover, each iteration of the diagnostic or treatment intervention has been proven, through rigorous clinical research, to do more good, and typically also less harm, than the previous iteration. As medical science evolves and new treatments are proposed, they are adopted into practice only after they have been proven in and of themselves to benefit patients and have also been proven to be an improvement on the current practice.

Take for instance information in the medical literature about a new drug to treat colon cancer. The drug would not be accepted into practice unless it were demonstrated, on balance, to be able to do more good and cause less harm in and of itself, and also were an improvement on the current drug being used. Good would be measured in terms of a longer cancer-free interval and increased length of overall survival, and harm would entail fewer side effects and complications. Even if the drug were validated to perform as it claimed, unless it also brought about more good and/or less harm than what was currently being used, it would not be adopted into clinical practice.

One of the ethical concerns for physicians is how, outside of a clinical trial, they are to know that the system and the physician collaboratively perform better than the physician performs independent of the system for a given decision.[12] From an ethics perspective, how can physicians know that the physician using the AI system to support clinical decisions will do more good and avoid more harm than the physician acting alone? A related question is that as the system continues to learn and thus change, how are physicians, in relying on the system to support their decisions, going to know that the system will continue to do more good and avoid more harm than the physician did without the system's support, or than the system and the physician did in their initial collaboration?[13]

In contrast to other medical interventions that are fixed, most AI systems used to support clinical decisions are continuously learning from those patients for whom the system provides decisional support. Take for instance a system that uses AI to reduce hospital mortality by identifying patients who will benefit from palliative care consultation based on a predicted survival of less than one year. The system will first be trained using medical information from a diverse set of patients with various disease processes who died within a year. The system will study patterns in the learning dataset so as to recognize those same patterns in future patients and identify those patients predicted to survive for one year or less. The patients for whom the system made a prediction will then be followed by the system to determine how accurate the system was in making a prediction. The system will then continue to hone its accuracy by learning from those patients for whom it has made a prediction. Before the system is deployed, it will be validated by comparing its predictions to those made by experts in the field.[14]

Subjecting AI decision support systems to clinical trials before their deployment is important in two ways. First, clinical trials assure that AI systems in collaboration with physicians are more beneficial and less harmful to patients than physicians acting alone.[15] In addition, prospective trials establish trust that AI systems have been held to the same ethical standard as other practices before they are adopted clinically. Using real-world data to train AI systems runs the risk of perpetuating the inequities, biases, and disparities that the data reflect. Conducting prospective trials to determine the clinical benefit of AI systems and auditing them for equity is one way of preventing AI systems from perpetuating the systemic injustices embedded in health care that are reflected in real-world data.

Similarly, auditing updates that are derived from the system's continuous learning will assure that as the system evolves, in concert with the physician, it does more good and avoids more harm than the physician acting alone and more than the system and the physician did in the previous iteration.[16] Just as a new version of a drug to treat colon cancer would be subject to a clinical trial before its adoption, even if its structure and mechanism of action were more akin to the original drug than not, so too would system updates need to be subject to the same kind of scrutiny.

Shared Professional Ethics between Physicians and AI Systems

A second consideration in comparing a physician's consulting a textbook or journal article to their collaborating with an AI system to make a clinical decision is that the physician can presume to share a set of professional ethics with the author of the textbook or journal article that they cannot similarly presume to share with the programmer of the AI system.[17] For instance, in adopting into clinical practice a new surgical technique described in a textbook, the physician can trust not only that the new technique has been proven to do more good and/or avoid more harm than the previous method, but also that other, adjacent ethical standards were maintained, such as that the personhood of those patients on whom the new technique was tested was respected and that their vulnerability as human subjects was protected. This is a professional ethical standard to which physicians and nonphysician researchers are held by their peers, government agencies, and licensing bodies.

Although it may well be true that the programmers and developers conform to some of the same standards as physicians and nonphysician researchers, because there are no similar widely held professional standards to which AI developers and programmers must adhere, it cannot be presumed. This is important in two ways. First, physicians cannot trust that, in the decision-making algorithm programmers have written, they have prioritized doing good and avoiding harm over other competing interests, such as increased revenue or efficiency.[18] In addition, it cannot be presumed that other values such as respect for patient autonomy, preservation of patient privacy, and truthfulness have been prioritized.

Take for instance a report by a physician or a nonphysician researcher about a new technique that diagnoses breast cancer earlier, more accurately, and with less risk of infection than the current practice of breast biopsy. Even though the physician who diagnoses and treats breast cancer realizes that the medical device company has other competing interests such as profit for its shareholders and market prominence for its brand, based on the fact that the research was conducted by and reported by physicians and medical researchers, the physician can be assured that at minimum the technique improves patient care, even if the device also makes shareholders a profit and the brand more prominent.

Certainly there have been physicians and medical researchers who have not abided by the high ethical standards their profession requires. However, these instances have been the exception rather than the rule, and overall reported medical research is trusted.

In contrast, when the physician receives a treatment decision from an AI system for a patient with lymphoma, how can the physician trust that the regimen the algorithm recommends has not been encoded to make certain trade-offs that would not have been made had the programmer prioritized patient care over decreased costs?[19] For instance, the system may have recommended a lower dose of a particular intravenous chemotherapy agent. The physician, receiving the decision, is unable to trust that the reason for a reduced dose is to avoid toxicity to the patient rather than to reduce cost for the payer.

Similarly, in identifying a patient predicted to develop type 2 diabetes, the physician cannot assume that the only interest encoded into the system is to allow for early intervention and reduced risk to the patient. The system may also share this information with other AI systems without the patient's knowledge. The patient may be profiled as someone likely to develop diabetes, thus affecting future clinical decisions. This would violate the patient's expectation of privacy and also potentially limit their autonomy.[20]

While these various outcomes can coexist and not be in conflict, when the physician turns to reports by medical researchers and other physicians to guide their decision-making about which chemotherapy agent to use to treat a patient with cancer, they can trust that the benefit of the patient was prioritized over other competing values, such as cost savings, and that other ethical priorities, such as patient autonomy and privacy, were upheld.

Those who develop and deploy AI systems can increase physician confidence that the decisions of the systems prioritize benefit, minimize risk, and do not violate other ethical commitments to patients in several important ways. Even in the absence of an explainable system, subjecting the system to a clinical trial and reporting the outcome of the trial in the medical literature would serve to build trust and to assure that ethical safeguards were observed.[21] Involving physicians, medical scientists, and bioethicists in the development, deployment, and auditing of the systems would also signal that members of the AI team could be assumed to share the same values and priorities as the physician caring for the patient who is receiving the system's clinical decision.[22] These strategies are currently used by medical device and drug companies, and they can serve to build trust in their products.

Prioritizing the Patient's Best Interest

A final ethical concern for physicians using AI to support clinical decisions, which differs from consulting a textbook or journal article, is the possibility that internal or external forces will compel the physician to accept the decision made by the system. There are several ways in which physicians may come to accept the AI system's decision without critical appraisal. Automation bias, as has been described in other disciplines utilizing AI for decisional support, may bias the physician to accept the system's decision based on the perceived superiority of the system to make predictions.[23] Additionally, physicians may become complacent, particularly after using the system and becoming familiar with it, essentially relinquishing their critical judgment to the system.[24] As the decisions that AI systems make become more explainable, meaning that there is some reasoning behind or association with the decision that is understandable to the physician, there is the risk that the explanation will seek not only to inform but to convince.[25]

In addition, payers may reimburse at higher rates, or only reimburse at all, if physicians enact the decisions made by the AI systems.[26] This is akin to the current practice of "preauthorization" in which payers will only pay for procedures that are approved through some often opaque process. There may be pressure from hospitals and health systems to comply with the AI system's prediction in order for the physician to be protected from peer review or quality assurance processes.[27] If, and when, AI decision support systems are accepted as the "standard of care," there may be legitimate concern that failure to enact the AI system's decision could result in grounds for malpractice claims.[28]

While some may argue that consulting an AI system in order to make a clinical decision is just like referring to a journal article or a textbook, as pointed out earlier, the important difference is that textbooks and journals give the physician information they can then use to make a decision. In contrast, an AI system makes a judgment and gives the physician a decision that, at least at this point in the technology, comes without an explanation or a reason. Combining the lack of explainability of AI with intrinsic or extrinsic pressure on the physician to accept the system's decision sets up an untenable situation wherein the physician loses their professional autonomy.

A defining part of being a physician is decision-making, and either unilateral implementation or engagement in shared decision-making with the patient. If a physician is unable to understand the AI system's decision and also limited in their ability to decide whether or not to implement the decision, that doctor in many ways is no longer in fact a physician. In this situation, the system assumes the essential role of the physician—the making and effective implementation of decisions—and the physician is the vehicle through which the system's decisions are implemented.

Shared Decision-Making and Paternalism

As AI is used to support clinical decisions, it will be important not only to preserve the physician's autonomy in making clinical recommendations to their patients but also to respect the value of shared decision-making between the patient and the physician. One risk is that traditional physician paternalism will be replaced by AI paternalism, in which the system knows best, even better than patients themselves.[29] Expecting a physician to "sell" the decisions made by an AI system to their patient betrays the patient's trust in their physician to make a recommendation which is in the patient's best interest. Requiring physicians to convince patients to accept the algorithm's prediction as a condition of reimbursement, regardless of whether or not the physician agrees, creates a conflict of interest in which the secondary interest, reimbursement, is at risk of influencing the primary interest, doing good for and avoiding harm to the patient. Even if the AI system acting in concert with the physician has been proven by clinical trials to do more good and avoid more harm than the physician acting alone, it is ultimately the patient's prerogative to determine the benefit of the "good," the burden of the "harm," and the acceptable balance of benefit and burden. AI clinical decision algorithms have been notably unsuccessful at incorporating patient values into their predictions.[30]

AI Clinical Decisions and the Standard of Care

Accepting AI systems as the "standard of care" has a different set of ethical considerations. *Standard of care* is a legal construct that can be used to determine whether the treatment provided to a patient by a physician is the same as a similar competent physician, of the same specialty, would provide under similar circumstances.[31] There is some variation state to state as to whether physicians are held accountable to the treatment a similarly competent *local* or a similarly competent *national* physician would provide.[32] If AI clinical decision tools were accepted as the *national* standard of care in a particular specialty, or for a particular set of clinical decisions, variation from the standard would be grounds for a malpractice claim if the patient was harmed as a result of the variance.[33] If the patient were harmed after the physician rejected the AI system's recommendation, believing their decision was correct and the AI system's decision was incorrect, the physician would likely be liable for a tort claim. Somewhat ironically, if the physician disagreed with the AI system's decision but went along with it as the standard of care, even if the patient were harmed, the physician would likely not be liable to a tort claim since the physician recommended a treatment aligned with the standard of care.[34] In this situation, even though the physician might not be subject to a malpractice claim, there would likely be damage to them personally and professionally because they had violated their ethical duty to do good and avoid harm by subjugating their judgment to the AI system. Moreover, ethical dilemmas would invariably arise in which the physician disagreed with the recommendation of the system and had to decide whether or not to go along with the system as the standard of care. If the physician espoused the system's recommendation, even if the patient were harmed, the physician would likely be protected. A good deal of moral courage would be required for the physician to act against the recommendation of the AI system.

RESOURCES FOR PHYSICIAN DECISION-MAKING: CLINICAL CALCULATORS

In an attempt to better predict patient outcomes, a number of clinical calculators and scoring systems have been developed. Some of these calculators and scores raise their own ethical concerns quite apart from those raised by AI CDS.[35] These calculators typically use retrospective data to predict the risk of some event based on patient characteristics.[36] For instance, the Eagle Score uses five clinical criteria (age >70 years, angina, diabetes, Q wave on EKG, history of congestive heart failure) to predict the risk of an adverse cardiac event following vascular surgery.[37] The scoring system is based on retrospective review of patients undergoing vascular surgery, using analysis of clinical factors, to predict the risk of an adverse cardiac event.

AI systems used to support clinical decisions, such as whether the patient's risk of a complication is too great for the patient to undergo surgery, differ from the clinical calculators being used currently in three ethically important ways. First, in order to receive the system's prediction, the patient must grant the system continuous access to their medical information, even after the prediction is made, as the system is continuously learning from the outcomes of the predictions it makes. Second, because the system is continuously learning from the predictions it makes about patients, the patient is a learning subject for the AI system. Finally, the decision about whether or not to undergo an operation should be based on shared decision-making between the physician and the patient. The process of shared decision-making involves the physician understanding and explaining the reasoning behind their recommendation.

The Patient as Donor to the AI System

Calculating a patient's Eagle Score doesn't require that the patient's data be given to an automated system. Typically, a physician can calculate a patient's score in their head and stratify a patient's risk while sitting in the examination room. In contrast, AI systems used to predict a patient's surgical risk will automatically pull data from the patient's EMR, use its algorithm to make a prediction, and then continue to access the patient's EMR to determine the accuracy of its prediction so that the system can continue to learn from the prediction.[38] In this way, the system's prediction is not a single isolated event but an ongoing connection between the system and the patient's EMR. The patient becomes a kind of donor, not of tissue or body fluid, but rather of their health information. Based on the physician's obligation to respect their patient as a person, a basic requirement is that patients be informed when they are donors, and that the donation be voluntary.[39] This places the physician caring for a patient in an ethical quandary, in that the AI systems used to make clinical decisions are often automated and out of the control of the individual clinician to opt in, or out, for an individual patient. As well, most physicians do not understand the implications of sharing a patient's EMR with a predictive algorithm well enough to be able to adequately explain it to the patient and get their consent.

Imagine a surgeon preparing to take a patient to the operating room to treat their colon cancer. An AI system has been deployed by the patient's hospital to automatically predict the risk of a postoperative wound infection for all patients undergoing surgery. Does the surgeon have an obligation to inform the patient that their medical record is being used to predict their risk of a postoperative wound infection, and that the system will continue to have access to their medical record? If so, how should the patient be informed? Should the surgeon discuss data protection and de-identification of data? Should the surgeon inform the patient that other data in their medical record may also be accessed? Should the surgeon inform the patient that even though their name and medical record number won't be attached to the data, the patient may nonetheless be profiled as a patient with colon cancer? Note that even though these questions are different from asking whether the patient should be informed if the system predicts that they have a higher risk of a postoperative wound infection, even though the surgeon can't explain to the patient why, the obligation to have either or both conversations has the same ethical foundation: the physician's respect for the patient's autonomy as a person.

The Patient as Subject for Learning

Similarly, when patients are used as subjects for learning, out of respect for their autonomy, physicians have an ethical obligation to inform the patient, and the patient's participation as a learning subject should be voluntary.[40] This applies not only when patients are participants in formal research studies, but also to when medical, nursing, and other professional students, and to a lesser extent resident physicians, participate in the patient's treatment. In some jurisdictions, such as New Jersey, the obligation to inform patients when they are learning subjects, especially for intimate exams, has been codified into law.[41] Similarly, there are examples from US case law of patients successfully bringing legal claims against physicians for not informing them when other medical professionals were involved in their treatment.[42] Just as with informing patients about the AI system accessing their electronic health record, physicians may also find themselves in a dilemma over whether and how to inform patients that they are learning subjects for AI systems.

Shared Decision-Making and the Duty to Inform

When the physician calculates a patient's Eagle Score prior to vascular surgery, particularly if the score suggests that the surgery would result in more harm than benefit to the patient, the surgeon has an obligation to explain to the patient why it is that surgery would carry such danger for the patient. The surgeon can reliably point to the experience of thousands of patients and say that when patients like the one sitting in the physician's exam room were taken to surgery, a known percentage had a heart attack, and of those, a known percentage died within thirty days. The physician can engage in shared decision-making with the patient by informing the patient and answering their questions.

A technically important, and ethically relevant, distinction needs to be made between AI systems that learn from labeled data and rules-based programming (supervised learning), and those that learn from unlabeled datasets and without following rules or algorithms (unsupervised learning). In the case of supervised learning, the AI system is trained on datasets that have been labeled to guide the system in predicting the desired outcome. These systems may also have rules that they must follow in making their predictions. It may be impossible to explain how the system made a specific prediction for an individual patient (ex post), but it is possible to cite the experts who annotated the datasets and the evidence-based clinical practice guidelines on which the system was programmed to make predictions (ex ante). In contrast, systems that are unsupervised are not given labeled datasets, nor are they provided with rules and algorithms on which to base their predictions. These systems make predictions using the historic data they are given, based on associations made entirely by the system, free of rules, labels, and guidance from evidence-based guidelines. These predictions are highly personalized to the individual patient, often very accurate, and unexplainable to the patient.[43]

Most AI systems currently being built for clinical decision-making are unsupervised. The power of these systems to make predictions is far greater than systems that are bound by rules and algorithms; however, both the physician collaborating with the system and the patient receiving the prediction do not know the reasoning behind the prediction or how it was made. This significantly limits the surgeon's ability to fully inform the patient and to engage in shared decision-making. For instance, a surgeon who is preparing to take a patient to the operating room for a colon resection consults with an AI system that predicts the risk of a serious postoperative wound infection. The system predicts a much higher than average risk of infection, which is relevant clinical information that should be shared with the patient. When asked why it is that the patient is at higher than expected risk, the surgeon may be able to point out common risk factors for infection after surgery: smoking, diabetes, poor surgical technique. However, the surgeon will be unable to tell the patient why the AI system made the prediction that it did for that individual patient, ex post explainability.[44]

This is in contrast to clinical calculators commonly used today to estimate surgical risk, such as the American College of Surgeons Risk Calculator.[45] These calculators are built using evidence-based clinical practices and with a limited set of variables that are known to affect surgical risk. The weight of the variables can be estimated so that patients can determine how to reduce the risk that a variable contributes. For unsupervised AI CDS, it will be unclear what factors the surgeon and the patient can alter so as to mitigate the predicted risk. Even if the AI system can predict the degree of certainty or uncertainty that an event will or will not happen, that is far from the robust model of shared decision-making that the physician is obligated to, that the patient deserves, and that is the foundation of professional trust.[46]

RESOURCES FOR PHYSICIAN DECISION-MAKING: COLLEAGUES

Physicians often collaborate with one another in taking care of patients. Generalists will consult with specialists who have specific knowledge of an organ system, such as the nervous or gastrointestinal system, or a set of disease processes, such as cancers and infections. The collaborations can take many different forms, from dropping into a colleague's office that is next door to one's own, to sending a patient for formal evaluation, to participating in a tumor board or case conference. Examining the nature of these collaborations and the responsibilities that accompany them sheds light on some of the ethical issues that will be encountered when physicians collaborate with AI systems to make clinical decisions affecting patients.

Physician-Physician and Physician-AI Consultations and Collaborations

When comparing collaboration with an AI system to consultation with a colleague to guide a clinical decision, four ethical concerns arise. First, how will the physician proceed in making a decision if there is disagreement between the physician and the AI system? Second, how will the physician present and explain the decision to the patient? Third, even if there is agreement between the system and the physician, the physician may incorrectly assume that it is because their reasoning was correct when it was not. This may lead to downstream harm to other patients based on the application of flawed reasoning. Finally, if the patient is harmed by the decision, whether or not there was disagreement between the system and the physician, who is morally responsible for the decision?

When consulting with a colleague in order to make a clinical decision, physicians often have a sense of the decision they would make but are looking for confirmation, or challenge, from a colleague. Using the example of consulting with a colleague to determine the optimal regimen to treat a patient with pancreatic cancer, there can be legitimate disagreement among clinicians about the optimal treatment regimen, combining varying chemotherapy agents and doses, radiation delivery methods and doses, and surgical approaches. If there is disagreement between the physician and a colleague, they can each present the reasoning behind their recommendations; they can critique each other's arguments; and they can refer to journals, texts, clinical calculators, and each one's experience, training, and education. If disagreement persists, a third colleague or a group of colleagues may be consulted. If the disagreement persists at this point, typically when it is unknown whether one regimen is superior to the other because they have not been subject to a clinical trial or research study, or when the regimens are regarded as clinically equivalent, the options along with the pros and cons would be presented to the patient. The important point is that the physician would know the relative benefits and harms of each regimen, why they are recommending a particular regimen, and why a colleague is recommending a different regimen.

In contrast, when there is disagreement between the AI system's decision and the decision that the physician arrived at on how to treat the patient with pancreatic cancer, the physician does not know why the system made its decision.[47] The physician cannot engage or reason with the system. Furthermore, the physician cannot know if the system's regimen will do more or less good or cause more or less harm than the regimen they decided on. It may well be that the physician's regimen will cause less harm but also provide less benefit to the patient than would the regimen the system decided on. The physician is then left to recommend one or the other regimen to the patient without knowing why the system made the decision that it made.

When there is disagreement between physicians, each physician can present their decision, along with their reasoning, to the patient. The patient can weigh the risks and benefits of each and apply their own values in choosing one or the other regimen. When there is disagreement between the physician and the

AI system over a chemotherapy regimen, the physician has no basis for explaining to the patient why the system made the recommendation that it made. The physician can only refer to the way the system was programmed, the power of its training, and perhaps the reputation of the experts who validated the system. In some cases, the physician may be able to point to the superior performance of physician-system collaboration for deciding on a chemotherapy regimen as compared to physicians making chemotherapy decisions alone, as proven by clinical trials. What the physician cannot do is present the relative benefits and harms of the system's decision as compared to their own decision and engage with the patient in shared decision-making to arrive at a conclusion. For instance, say the regimen the system recommends has more toxicity for that patient than the regimen the physician recommends. However, the system's regimen also has better predicted survival for that patient than the regimen the physician recommends. If the physician knew this about each regimen, they could present the relative risks and benefits to the patient and, based on the patient's values, the patient could decide, after considering the relative weight of each, whether increased toxicity was an acceptable trade-off for improved survival.

If the decision of the system is unexplainable, or if the explanation is such that its risks and benefits cannot be effectively communicated to the patient, the physician is left to either disregard the system's decision and only present the patient with their own decision, to disregard their decision and present the patient with the system's decision, or to present both decisions to the patient without being able to truly engage with the patient in shared decision-making, which is one of the core duties of the physician.

Not only can the physician not know the relative benefits and harms of the system's recommendation, they also cannot know whether the system prioritized other competing values, such as cost savings and operational efficiency, over optimizing benefit and minimizing harm to the patient.[48] Finally, the physician cannot know whether the system is upholding other ethical obligations that the physician owes to the patient. One example is whether the system respects the patient's privacy by not sharing the patient's health information across platforms. Another is whether the patient's autonomy is upheld by not profiling the patient as someone with pancreatic cancer and thus impacting and possibly restricting future decisions made by the AI system.[49]

While an explanation for the system's decision is especially important if the decision differs from that which the physician arrived at on their own, if the system and the physician agree, the physician may be led to assume that the reasoning they used to arrive at the clinical decision was accurate. This assumption may not be true, and the physician may base other clinical decisions on a set of assumptions they believe to be true based on previous agreement with the AI system, but which are not true, potentially leading to patient harm.

The Physician's Moral Responsibility

Ultimately, the physician must choose how to proceed. From a legal perspective, physicians are held liable based on whether or not their decision was consistent with the standard of care.[50] If harm occurs to a patient and the decision made by the physician is determined to be consistent with the standard of care, the physician would generally not be held liable. Legal theorists argue that if and when AI systems are adopted as the standard of care, physicians may be held liable if they reject the system's decision and the patient is harmed.[51]

Putting aside the consideration of legal liability, disagreements between the physician and the AI system have the potential to be ethically and morally distressing for the physician. For most physicians, doing good and avoiding harm to patients is not only a professional ethical responsibility but also part of their personal morality. A physician who has to choose between their own judgment and that of the system, without a basis other than the system's epistemic authority, faces a huge moral burden.[52] The burden is there even if the patient is not harmed. Whether recommending their own treatment regimen to the patient and discounting the recommendation of the AI system or vice versa, the lurking question remains: Could the patient have been helped more and harmed less by the decision that was not enacted than by the decision that was enacted?

When there are disagreements between the physician and the system, there is a potential that physicians may come to distrust their own judgment as well as that of the system. If the physician enacts their own decision, contrary to the system, and the patient is harmed, not only is the physician morally and also likely legally responsible, the physician may also come to distrust their own judgment.[53] Similarly, trust in their own decision-making may be eroded if the physician enacts the system's decision over their own only to realize that had they enacted their own decision, the patient would have been harmed. If the physician enacts their own decision, contrary to the system, and discovers afterward that had they enacted the system's decision, the patient would have been harmed, the physician may come to distrust the system's decisions. Similarly, if the physician enacts the system's decisions over their own and the patient is harmed, the physician may also come to distrust the system.

In addition to these concerns, physicians also face a potential loss of professional autonomy as AI systems for clinical decisions are accepted as the standard of care, as payers base reimbursement on whether or not the decision of the system was followed, and as health systems review the quality of a physician's performance against what was predicted by AI systems.[54] While it is unclear how the limitations on professional autonomy will affect the physician's legal responsibility for decisions made by the system, it is more clear how the loss of autonomy is likely to lead to decreased physician well-being and overall professional satisfaction.

THE QUESTION OF AI-PHYSICIAN COLLABORATION

The purpose of comparing here the ways in which physicians currently make decisions to their collaboration with AI CDS is not only to raise ethical concerns that will result from that collaboration but also to better understand the nature of the collaboration. Although there are features of the physician's use of journals, texts, and clinical calculators in the use of AI CDS, the physician's collaboration with AI is most like their collaboration with a colleague, primarily in that like a colleague, the AI system provides a judgment in the form of a clinical decision, just as a colleague would. As has been discussed, there are some significant limitations to the AI systems compared to a colleague, mostly the inability of the system's decision to be explained and of the system to be reasoned with.

Two other differences, which have been alluded to, are also very important for understanding the nature of the collaboration. First, typically if a physician wishes to consult with a colleague, the collaboration is freely requested and voluntarily given. In contrast, the AI CDS will be inserted into the EMR and its decision given without the physician requesting or perhaps even wanting it. In addition, unlike most colleagues, who are peers, the system is not a peer. In fact, the decision of the system may carry more weight than would a colleague's decision and may even be expected to carry more weight than would the physician's own decision.

The system thus becomes a third party in the physician-patient relationship, without moral or legal responsibility, without the professional ethical duties that attend the physician, and potentially with the expectation that more weight will be given to its clinical decisions than the physician's. Physicians are thought of as having agency for the decisions they make and either enact unilaterally on behalf of the patient or present to the patient in order to engage in shared decision-making. The physician's agency is derived from the act, making a decision, and the intention, to do good and avoid harm to the patient. As AI CDS systems do not have intentions in the same way that physicians, and for that matter, other humans do, they lack complete agency in and of themselves. However, the systems do perform an action, in that they make a decision, and the decision is programmed toward some end point, such as increased survival or cost savings. As the physician enters into collaboration with the system, they seem to share their own agency with it and thus provide the system with actualization of the system's partial agency.[55] The decision made by an AI CDS system in a vacuum would lack full agency, but upon introduction into the physician-patient relationship as a third entity, the system derives agency by sharing in the physician's agency.[56]

SUMMARY AND CONCLUSION

The patient is at the center of the physician-patient relationship, and the physician's core ethical duty is to do good and avoid harm to their patients through the decisions they make. Comparing the ways physicians currently make decisions to the use of AI CDS raises ethical concerns, many of which are related to the lack of explainability of these systems. Compared to textbooks and journals, AI systems make a judgment and deliver a decision. Whereas the information reported in journals and texts has been demonstrated through clinical research to benefit patients, most AI CDS systems have not been subject to research demonstrating that the system sharing in decision-making with the physician does more good and less harm than does the physician acting independently. Moreover, as many AI CDS systems are continuously learning, it cannot be assumed that they will continue to do good and avoid harm as they evolve. Furthermore, there is no reasonable expectation that these systems will support other ethical obligations that physicians owe to their patients, such as privacy and respect for autonomy. Finally, as these systems are deployed, physicians may subjugate their own judgment to that of the system, through automation bias, payer or health system expectations, or fear of legal liability if the systems are accepted as the standard of care.

When comparing the use of AI CDS to clinical calculators, ethical concerns about donating their health information to the system and serving as subjects for the system's learning pertain more directly to patients. Physicians have a duty to inform patients when they are donors and when they are learning subjects, explain the associated risks and benefits, and ensure that patients' participation is voluntary. This places an ethical onus on physicians, particularly since many physicians will not fully understand how these systems work or the associated risks. Unlike when they use clinical calculators, physicians will be unable to explain to patients the reason behind the system's prediction and thus unable to fulfill their obligation to participate in shared decision-making with patients.

Finally, unlike consulting a peer, if there is disagreement between the system's decision and the physician's decision, the physician cannot understand the system's reasoning, nor can the system understand the physician's reasoning. The physician and the system cannot critique each other or engage in back and forth until a decision is reached. Physicians can trust that in the decisions they make, their colleagues will prioritize doing good and avoiding harm, respect the patient's autonomy, and maintain privacy. AI CDS systems do not have similar professional obligations, and physicians cannot know that these principles are upheld by the system. As has been seen with other comparisons, unlike when they consult with a colleague, physicians will have no basis for presenting the system's decision to the patient. Even if the physician and the system agree, it may be for different reasons, and the physician may incorrectly assume that the two agree because the physician's reasoning was accurate, whereas it may not have been. The physician may go on to apply the same reasoning to other, similar clinical situations, potentially resulting in harm. The physician would not likely be held legally responsible if the decision was in accord with the standard of care, which might be the AI CDS, but the physician might still feel morally responsible. In addition, disagreements between the physician and the system, whether or not the patient was harmed, can lead to physicians distrusting themselves and the system.

It is of paramount importance that these ethical concerns be addressed before the systems are widely deployed in clinical settings. Some of these concerns could be addressed by "explainable AI." However, it is not clear whether the aim would be understanding or complying with the system's decision and whether or not physicians would trust the explanation. Other concerns could be allayed by subjecting the systems to clinical trials before they are deployed and continuously auditing them to be certain that as they learn they also continue to do more good and less harm than their previous iteration, that patient autonomy and privacy are maintained, and that patients are informed about the use of AI CDS in their care. Finally, there should be a clear method for resolving disagreements between the system and the physician, and the physician's professional autonomy should be respected. In this way, these systems have the greatest likelihood of living up to the promise of improving patient care, reducing patient harm, and making health care more affordable and accessible.

CASES

Case Study 2.1: Physician Liability in Opposing AI

Ms. Smith is a forty-six-year-old woman who presents to the emergency department with severe epigastric pain and nausea and is diagnosed with severe biliary pancreatitis, a condition in which stones and sludge from the gallbladder travel down the common bile duct—the tube that drains bile from the liver and gallbladder into the intestines—and lodge in the region where the common bile duct is joined by the pancreatic duct—the tube that drains digestive enzymes from the pancreas into the intestines. Blockage of the pancreatic duct causes the digestive enzymes to leak into surrounding tissues and organs, liquifying their proteins and destroying their structural integrity. The health-care team's duty is to support Ms. Smith's organ systems while surveilling for and treating the complications of pancreatitis—primarily bleeding and infection—if they arise.

The decision-making process for managing severe pancreatitis is complex. Dr. Silen, Ms. Smith's physician, must choose which tests to order in surveilling for bleeding and infection; interpret the test results; and decide if and when antibiotic therapy should be initiated, whether one or more of many invasive procedures should be performed to control infection or bleeding, when and how to deliver nutrition to allow Ms. Smith's body to heal and rebuild, and how much fluid volume should be administered through intravenous catheters to keep Ms. Smith from becoming dehydrated without making her body swollen with excess water. Excess water can flood the lungs, causing respiratory failure and requiring placement of a breathing tube for mechanical ventilation, while dehydration can cause kidney injury and failure, which can require initiation of hemodialysis. Pancreatitis-induced inflammation makes Ms. Smith's blood vessels leaky, so most fluid administered through intravenous catheters escapes from her bloodstream and worsens swelling. Before developing biliary pancreatitis, Ms. Smith suffered from chronic kidney disease, chronic obstructive pulmonary disease, and coronary artery disease, so her risk of kidney failure and heart attack are high, especially if she becomes dehydrated, and her risk of respiratory failure is high, especially if her lungs become swollen with excess water. Unnecessary antibiotics can cause side effects and overgrowth of antibiotic-resistant super bacteria; a delay in administering necessary antibiotics is associated with increased risk for death. Similarly, unnecessary invasive procedures for infection or bleeding can cause further infection, bleeding, and injury to nearby organs, and failure to perform necessary procedures to control infection or bleeding is associated with increased risk for death. Dr. Silen and her team must walk a decision-making tightrope, knowing that an errant decision can have severe consequences.

Given these major decision-making challenges in managing severe pancreatitis, Dr. Silen compares her clinical reasoning and intuition with recommendations from a CDS system built around a reinforcement learning algorithm that identifies discrete actions at discrete timepoints that are associated with optimal long-term outcomes. When the CDS recommends the action that Dr. Silen has already decided upon independently, she proceeds as usual; when it recommends a different or opposite action, Dr. Silen carefully reexamines the decision and tries to understand why the model disagrees with her. Most times, she knows something the model doesn't (e.g., information garnered from examining the patient and speaking with family members). Sometimes Dr. Silen realizes that she was overlooking a complex pattern emerging from interactions between several test results and observations; in such cases, she accepts the model recommendations. Using this framework, Dr. Silen and her team make frequent, careful deliberations, and are cautiously optimistic when Ms. Smith responds well to initial resuscitation and appears to stabilize after a single episode of low blood pressure during her first night in the intensive care unit (ICU).

On her third day in the ICU, Ms. Smith develops recurrent, worsening, low blood pressure. Her vigilant ICU nurse notices flank bruising outside the area in which he administers subcutaneous medications that prevent blood clot formation (in which bruising normally occurs), and calls Dr. Silen to the bedside. She recognizes the bruising as a Grey Turner sign, indicating that internal bleeding likely has occurred from pancreatic enzymes eroding through a blood vessel within the past few days and suggesting that recurrent hypotension represents rebleeding from clot dislodgement. The AI-enabled CDS takes in new vital sign and laboratory measurements as model inputs and generates updated recommendations. Ms. Smith's hematocrit—a laboratory measurement of red blood cell concentration—remains normal, because changes in blood concentration will occur soon but have not occurred yet. The CDS system, blind to the Grey Turner sign and falsely reassured by the normal initial hematocrit level, associates worsening hypotension with worsening infection and recommends initiation of powerful antibiotics and placement of a drainage tube through the abdominal wall into a fluid collection around the pancreas. Dr. Silen wisely acts against the recommendation of the algorithm and sends Ms. Smith for an angioembolization procedure, in which the bleeding blood vessel will be identified and plugged under X-ray guidance. While being transported from the ICU to the Interventional Radiology suite, Ms. Smith suffers a cardiac arrest and cannot be resuscitated. Her devastated family defers postmortem examination.

Ms. Smith's case is reviewed by a panel of physicians to determine whether there were opportunities for improvement. It is noted that in the minutes leading up to Ms. Smith's demise, the CDS system predicted a less than 20 percent probability of in-hospital mortality and made recommendations that were contrary to Dr. Silen's actions. Based on review of the electronic health record and according to a highly accurate reinforcement learning algorithm, the contents of which will be available during legal recourse, it appears that Dr. Silen deviated from recommendations, and the patient suffered an unanticipated mortality.

Without proof that hypotension was due to bleeding, Dr. Silen has only a rare and occasionally unreliable physical exam finding to support her decision to disagree with a highly accurate CDS system.

QUESTIONS

What are the risks and benefits of using AI-enabled CDS predictions and recommendations in legal proceedings regarding medical malpractice?

Should AI-enabled CDS allow manual inputs to steer predictions and recommendations toward human intuition or information that is inaccessible to the model? What are the potential dangers of building human intuition into CDS?

Case Study 2.2: Physician Replacement by AI

Dr. Obslo is a radiologist who works for a large hospital corporation. Since completing residency ten years ago at the top of his class, he has dutifully interpreted X-rays, computed tomography (CT) scans, and magnetic resonance imaging (MRI) by carefully applying hard-won skill in identifying normal and abnormal imaging findings and communicating those findings to the physician who ordered the test by written and verbal reports. Dr. Obslo has always felt a special sense of purpose when patients' diagnoses were discovered or confirmed by his interpretations.

After eleven years in practice, computer vision AI models disrupted Dr. Obslo's workflows. A growing body of evidence demonstrated that AI could interpret commonly performed imaging studies with accuracy equal to that of board-certified radiologists. The administrators running Dr. Obslo's hospital corporation made a sound business decision to invest in computers that require no salary or benefits; work 24 hours per day, 7 days per week, and 365 days per year; and can perform in seconds the same volume of work that a radiologist might perform during an entire week. Therefore, Dr. Obslo's role evolved to focus on rarely performed imaging studies for which AI prediction models were not available and to spot-check interpretations generated by the AI system.

Dr. Obslo's sense of purpose in discovering or confirming patients' diagnoses begins to subside, and his work satisfaction and overall well-being follow suit. Despite having less work to do, he makes errors more frequently and with greater severity. His supervisors relieve him of the responsibility to spot-check interpretations generated by the AI system. As more images are banked in the system, there are few image types for which the volume of data is insufficient to train the AI system. Human interpretation of those remaining rare image types is consolidated among an ever-diminishing group of radiologists, from which Dr. Obslo is subsequently cut. Despondent and jobless at age forty-two, he wonders how four years of college, four years of medical school, and five years of residency have left him with an unemployable skill set.

QUESTIONS

Do hospital administrators have a moral responsibility to employ radiologists rather than AI systems, under the condition that their performance is equal?

What are the potential long-term consequences of displacing human workers?

Case Study 2.3: AI-Enabled Physician Freedom from Rote Tasks

Dr. Artemis is a pathologist who works for a large hospital corporation. Since completing residency ten years ago at the top of her class, she has dutifully interpreted gross and microscopic biopsy and organ specimens by carefully applying hard-won skill in identifying normal and abnormal imaging findings and communicating those findings to the physician who ordered the test by written and verbal reports. Dr. Artemis has always felt a special sense of purpose when patients' diagnoses were discovered or confirmed by her interpretations.

After eleven years in practice, computer vision AI models disrupted Dr. Artemis's workflows. A growing body of evidence demonstrated that AI could interpret commonly performed imaging studies with accuracy equal to that of board-certified pathologists. The administrators running Dr. Artemis's hospital corporation made a sound business decision to invest in computers that require no salary or benefits; work 24 hours per day, 7 days per week, and 365 days per year; and can perform in seconds the same volume of work that a pathologist might perform during an entire week. Therefore, Dr. Artemis's role evolved to focus on rarely performed imaging studies for which AI prediction models were not available and to spot-check interpretations generated by the AI system.

Dr. Artemis sought and received an additional role. Since graduating from medical school, her personal contact with physicians outside pathology had been limited, and her personal contact with patients was nonexistent. She now recognized that with AI doing the rote work of interpreting biopsy and organ specimens, she could broaden and deepen her contact with the physicians who obtained the specimens and might benefit from a physician-to-physician conversation about the findings. Perhaps more importantly, she could initiate personal communications with patients anxiously awaiting information about their diagnosis: Was it cancer? Had it spread to the lymph nodes? Did we find the cellular markers that indicate that the tumor is sensitive to a promising new medication that uses the patient's own immune system to fight the cancer? Dr. Artemis relished these opportunities, and her colleagues and patients were grateful for the warm, caring demeanor with which she shared her wisdom. She was promoted to director of pathology services and appointed several trusted colleagues to serve as associate directors of physician communications and patient communications. The ranks of her department quickly filled with intrepid pathologists who approached their work with a sense of creativity and altruism that cannot be encoded with AI.

QUESTIONS

What policies and work structures can facilitate the evolution of human work in health care toward tasks that are best performed by humans?

In addition to personal communications, what kind of tasks might be performed by radiologists or pathologists when they are no longer bound to rote, commonly performed tasks?

CHAPTER THREE

The Patient and AI Clinical Decision Support Systems

THE PATIENT AND THE PATIENT-PHYSICIAN RELATIONSHIP

As medicine becomes increasingly complex, it is easy to forget that human beings, whom we call patients, are at the center of all that we do. The word *patient* is derived from the Latin *patiens*, which means suffering. Thus, the aim of medicine is to relieve human suffering. When developing, programming, validating, and deploying AI systems for clinical decision-making, the benefit of the patient should be the primary goal.

This chapter begins by exploring the perspective of the patient in the physician-patient relationship and what it means to the patient that AI CDS is introduced into the relationship through the physician's collaboration. Trust, which is at the center of the patient-physician relationship and is necessary for a therapeutic alliance to be established between them, is established and cultivated through specific actions that the patient rightfully expects from their physician: to inform them, to make their participation voluntary, to maintain confidentiality, to prioritize their well-being, to avoid conflicts of interest and avoid bias, to not abandon them, and to be truthful and accountable.

It is largely through fulfilling these expectations that trusting relationships between physicians and patients are established. It is important to reflect on how introducing AI CDS into the physician-patient relationship might impact patient expectations and the bond of trust between patients and physicians. Doing so will highlight some of the ethical concerns applicable to patients, not only to describe the concerns but also to attempt to address them, so that AI CDS can be introduced in such a way as to foster rather than rupture trust in the therapeutic relationship.

THE RELATIONSHIP BETWEEN THE PATIENT, THE PHYSICIAN, AND AI CDS

It is important to recognize that patients place trust in their physicians, usually in the setting of uncertainty, often regarding actions and decisions with major impacts on quality of life and occasionally regarding life or death situations. This act of trust allows a therapeutic relationship to form between the patient and the physician. It is within this trusting, therapeutic relationship, which is framed by clinical uncertainty, that medical decisions are made. While the overall goal of introducing AI CDS into the relationship is ideally to decrease clinical uncertainty and thereby increase trust, the fact remains that the primary trusting relationship is between the human patient and the human physician. The AI CDS is introduced into the relationship through the physician in a form of shared agency, a human-machine collaboration for the purpose of making medical decisions, built on the central hypothesis that humans and intelligent machines working together will outperform humans alone. AI CDS research occasionally compares humans to machines in decision-making processes, but that contest is less relevant in clinical settings and is not the focus of this text.

The means by which patients interface with AI CDS differ from many other human-AI interactions in ethically relevant ways. First, and perhaps most important, is that the system's interface with the patient requires two relationships: one between the physician and the patient and one between the system and the physician. The physician is an intermediary between the system and the patient, though the patient is the ultimate recipient of the system's action. To understand the relevance of the indirect relationship between the patient and system, it is useful to first examine the process by which medical decisions are typically made. For instance, when treatment decisions are made, physicians first consider all possible treatments for a given illness and then narrow the possibilities based on what they believe will most likely help the patient achieve their desired outcome. That is the first clinical decision. Next, the physician may present the treatment options to the patient and engage with the patient in shared decision-making in order to determine which treatment the patient finds most acceptable. This is the second decision, and it is the focus of this section.

Take for example a patient who presents with high blood pressure. There are hundreds of possible treatments for high blood pressure, ranging from diet and exercise to sometimes complex combinations of medications. The physician considers the range of possibilities within the context of the patient's medical information and recommends a treatment regimen. Every treatment regimen has its benefits and burdens, and as we have seen in chapter 2, the physician is guided in their decision-making by recommending the treatment that will do the most good and the least harm. The physician then presents their recommendation to the patient, describes the benefits and burdens, and engages in a process of shared decision-making with the patient whereby they together decide upon a course of treatment.

When AI CDS is introduced into the relationship, the physician will first collaborate with the system to arrive at a treatment recommendation, the first decision, and then present the recommendation to the patient to make the second decision. The patient, who is the final recipient of the system's prediction, has an indirect relationship with the system, the physician has the primary relationship and thus is also the intermediary. In contrast, when Betty asks an AI-powered voice recognition application to call her mother, there is a single relationship in which Betty both directly collaborates with the AI system and is also the recipient of the system's action. Relevant to this discussion is that because the primary relationship is between Betty and the system, Betty knows that the system exists, has on some level been informed of the benefits and burdens of using the system, and therefore has a basis for deciding whether or not to trust the system to call her mother.

When introducing AI CDS into the physician-patient relationship, patients may not have been informed about the collaboration between their physician and the system, including the extent of the collaboration; may not have been informed of the benefits and burdens of using the system to make their medical decisions; and may not have a basis for trusting the collaboration between their physician and the system. As AI CDS is introduced into clinical practice, physicians will need to determine patient preferences about the use of such systems and help patients weigh the risks and benefits of their physician collaborating with an AI system. Doing so will build trust between the patient and the physician, which is essential to the therapeutic relationship and is protected by a code of ethics.[1]

TRUST AND THE PATIENT-PHYSICIAN RELATIONSHIP

In order to explore the effects that introducing AI CDS into the physician-patient relationship may have on the patient's trust in their physician, it is helpful to consider what patients expect of their physicians in order to gain their trust. First, trust is fostered when physicians inform their patients about their diagnosis, treatment recommendation, or prognosis and engage with them in some form of shared decision-making.[2] Usually this involves both the physician informing the patient and the patient giving their consent to proceed in enacting whatever decision was made.

Informing the Triune Patient

Besides being informed about their diagnosis, treatment, and prognosis, trust is also built when physicians inform patients when they are donating tissue, organs, or genetic material; when patients are being used as learning subjects; and when patients are research subjects. Patients commonly occupy these other roles as donors, learning subjects, and research subjects. However, these roles are distinct from the role of patient qua patient. These other roles have their own ethical obligations and expectations.[3] Most importantly for this discussion, it can not be assumed that patients are also necessarily also donors, learning, and research subjects. Furthermore, it is essential to understand that the provision of medical care to patients can never be contingent on

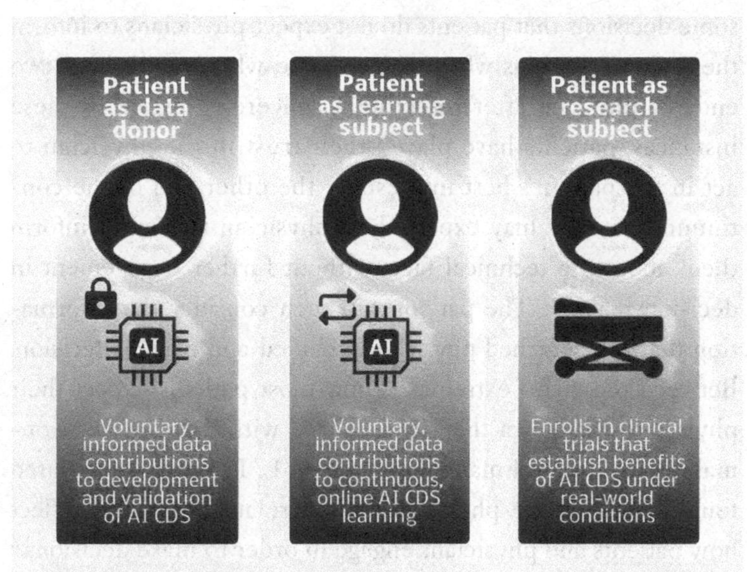

Figure 3. The triune patient

patients' participation as donors, learning, or research subjects.[4] These other roles, though occupied by the same person, are distinct. Taking on these other roles by the patient must be voluntary, informed, and explicitly consented to. History is rife with examples of patients not being informed when they were being used as donors and learning and research subjects, and the irreparable damage to trust in the physician-patient relationship that resulted.[5]

Informing the Patient as Patient in Medical Decision-Making

There is a continuum of what patients expect their physicians to inform them about. At one end of the continuum, there are some decisions that patients do not expect physicians to inform them about, such as what suture to use when reattaching two ends of intestine after resecting a cancerous lesion. In these instances, patients have placed their trust in the physician to act in the patient's best interest. At the other end of the continuum, patients may expect their physician to simply inform them about the technical facts without further engagement in decision-making. The patient will then consider the information that the detached physician provided and make a decision. Between these two extremes is how most patients expect their physicians to inform them and engage with them in decision-making. Ezekiel J. Emanuel and Linda L. Emanuel articulated four models of the physician-patient relationship that reflect how patients and physicians engage in order to make decisions.[6]

Most patients expect at least three things when engaging with their physician to make a decision. First, patients expect to receive information regarding risks, benefits, and alternatives to whatever the physician is recommending. Patients also expect the physician to explain to them how the physician arrived at their decision to recommend a particular course of action or a range of options over other available options. Finally, patients expect physicians to respect their values and preferences when making medical decisions. Medical decision-making is the essential action of the physician, whereas informing and engaging in decision-making with the patient is the essential ethical obligation of the physician. These go hand in hand.

When a physician has collaborated with an AI system in order to make a clinical decision for a patient, the physician should be able to explain not only the risks and benefits of that treatment, but also those of AI-enabled decision support (i.e., the potential for errant or biased recommendations) and, using lay terms, how the AI algorithm arrived at its recommendation. However, in many cases, because the system's decisions are not explainable, the physician will not be able to inform the patient why that treatment will be specifically beneficial to that patient, and why that particular treatment was recommended over other treatments.

For instance, following severe trauma to an extremity, it is often clinically challenging to predict which patients should undergo immediate amputation to facilitate rehabilitation and adoption of a prosthesis for function, and which patients should undergo surgery aimed at reconstruction and restored function of the traumatized extremity. Patients neither want to amputate an extremity that otherwise could be saved nor, in most instances, to undergo multiple, painful procedures; likely complications including infection and tissue loss; and an extended hospital stay, only to have the extremity ultimately be amputated. An AI CDS system that predicts, with a high level of accuracy, which patients should undergo immediate amputation and which should be treated in order to save the limb has the potential to be very beneficial to patients and physicians. Such a system would also reduce hospital costs accrued from treatments on an extremity that would ultimately be amputated.

However, a patient who has been informed that an AI system has predicted one course of action over another, without being able to understand why that prediction was made for that patient, may well feel as though they do not have enough information on which to apply their own values to the AI-enabled recommendation so as to arrive at a final decision. The patient may also feel some degree of frustration and anger that such a high-stakes decision lacks concrete guidance. These feelings, if experienced, may be displaced onto the physician, since it is the physician who has the relationship with the patient. The patient may feel further vulnerability and anger if they learn that their payer will only reimburse treatments that are aligned with the system's prediction. The patient's physician may also have to deliver this message, leading to further stress on the relationship between patient and physician. The patient may begin to distrust the physician, perceiving that the physician is not being their advocate or acting in their best interest. The patient may also come to distrust the physician because the patient believes that the physician's primary interest is not the patient but rather their own monetary interest or the financial interest of the health system.

Building Trust: Explainability and Transparency

Building trust raises two ethical issues relevant to the patient. First, because the patient will not know the benefits that the treatment will provide for them specifically, or why other alternative treatments may not be as beneficial, they will not have a basis on which to weigh the options fully and apply their own values. This leaves the patient at a significant disadvantage for making a fully informed decision about their medical care. Physicians are justified in limiting the treatment options presented to patients based on what is beneficial and what is achievable.[7] There is no obligation to list every possible treatment. However, it is important that physicians explain to patients their reasons for choosing the options they are presenting, particularly the benefit and the risk of each option relative to the individual patient. In contrast, AI systems will offer patients a single decision that is based on the desired outcome the system was programmed to predict. Because the prediction is unexplainable, patients will not be able to weigh the prediction's benefit relative to the risk. Additionally, the outcome that the system is programmed to predict may not be congruent with a benefit that the patient values.

A second ethical concern is that physicians may try to fill this void by explaining in general why the treatment is beneficial or speculating about why it is specifically beneficial to that patient, even though the physician's reasoning may not actually be the basis for the system's prediction.

Such generalization and speculation may mislead the patient. In speculating and generalizing, it may not be clear to patients that their physicians have collaborated with an AI system to make a clinical decision that the physician is in turn recommending to the patient. In order to maintain patient trust, physicians may speculate and generalize, particularly if they believe that patients would find their collaboration with an AI system unacceptable; if the patient would refuse the collaboration; or if the physician doesn't want to admit that they are under medical-legal, health system, or payer pressure to accept the system's decision. Physicians may also think that it is not important to disclose to the patient that they collaborated with an AI system. However, as long as the predictions made by these systems are not understandable by the physician and explainable at the patient level, it would seem that physicians have two options. They can either speculate and generalize without revealing the collaboration with the AI system, or they can admit that they collaborated with an AI system to make the clinical decision, and that although the physician can speculate and generalize, they can not know for sure why that decision was made for that patient. The only ethically justifiable exceptions to the expectation that physicians would inform patients about their collaboration with an AI system are if the patient has expressed their preference not to be informed, or if the physician feels unprepared to explain the collaboration and believes that an attempt at explaining it would be more harmful than beneficial. The latter exception could only be invoked under very limited circumstances and would necessarily lead the physician to better prepare in the future to explain their AI CDS collaboration to their patients.

As AI systems become more sophisticated, their predictions may also become more understandable by physicians and explainable to patients. While this may address the concern that the specific reason for a particular recommendation has not been explained to the patient such that a fully informed decision can be made, it has the potential to deepen the concern about withholding from the patient the collaboration between the system and their physician. If physicians receive from the AI system both the clinical prediction and the underlying reasoning, and it coincides with the physician's own decision and reasoning, physicians may believe that there is no need to inform the patient about the collaboration.

Withholding from the Patient the Physician-AI Collaboration

What could justify withholding the collaboration between the system and the physician? The physician may believe that the decision and reasoning is their own, that the system has simply confirmed what is theirs, and that the contribution of the system to the decision is so inconsequential that revealing the collaboration is unnecessary. The physician may also believe that revealing the collaboration would needlessly confuse or overwhelm the patient, or the physician might assume that the patient doesn't care about the collaboration.[8] While all of these facts may be true, they do not justify withholding the collaboration from the patient.

If the physician believes that agreement between them and the system justifies withholding information about their collaboration with the system from the patient, then should the patient only be informed of the collaboration if there is disagreement between the physician and the system? If patients are only informed in instances of disagreement, they will not have a representative sample of occasions of collaboration between physicians and AI CDS. This may lead patients to believe that collaborations between physicians and AI systems only lead to disagreements. Depending on the patient's perspective, they may come to either place unjustified trust in the physician's decision as being always correct and thus discount the AI system or place unjustified trust in the AI system's decision and discount their physician.

Transparency about the collaboration between the AI system and the physician whether there is agreement or disagreement enables the patient to have a balanced perspective about the collaboration, which will more than likely be in agreement rather than disagreement. Also, if patients are only informed when there is disagreement between the system and the physician, they may wonder whether or not there was collaboration between the physician and the AI system when making other medical decisions.

Knowing that there is potential for collaboration between physicians and AI systems for some decisions, if the collaboration is not transparent, patients may ask their physician whether or not the physician collaborated with an AI system in making their own medical decision. If there has been collaboration, and it was not revealed to the patient from the beginning, the patient may lose confidence in the physician's honesty and trustworthiness.

Withholding the collaboration between the physician and the system because informing the patient would be confusing or overwhelming, or because it is assumed that most patients won't care about the collaboration, fails to show respect for patient autonomy and generalizes the patient's experience. While many patients may not care about the collaboration, some will find it important, and neither condition should be assumed. In general, the strongest justification for withholding information from a patient is that the patient has told the physician that they do not want to know a particular piece of information.

One could argue that the physician has already made the trust calculus between themself and the system, and the patient has chosen to trust the physician and their decision-making process, and thus by a transitive property, the patient will trust the AI CDS. However, a patient's trust in their physician is continually reinforced by the process of informing the patient and sharing in decision-making.[9] From a consequentialist perspective, one could argue that if the physician withheld their collaboration with the AI system from the patient, the patient was harmed because of the collaboration, and the patient later discovered the collaboration, the patient may reasonably conclude that the physician is dishonest.[10] This could lead to further harm because of damage to the physician-patient relationship. More generally, if some physicians inform patients about their collaboration, and other physicians do not, patients may come to distrust those physicians who do not disclose their collaboration.

Respect for Patient Values in Medical Decision-Making

Besides expecting to be informed about the risks, benefits, and alternatives to a clinical decision and the reasoning behind the decision, patients also expect to have their values taken into account when physicians decide on a course of action. In terms of medical decision-making, as has been explained, the physician decides on a course of action to recommend to the patient; the physician informs the patient of their recommendation and the risks, benefits, alternatives, and reasoning behind the recommendation; and then the physician and the patient engage in the process of decision-making. A patient's values deeply influence not only the intended outcome but also what things are acceptable and unacceptable in achieving that outcome. For instance, some patients place value on life itself, even if they are unable to interact meaningfully with others or their environment because of neurologic damage, and choose, often through advance directives, to use artificial hydration and nutrition to maintain life. In contrast, other patients choose to forego artificial hydration and nutrition if they are not able to enjoy a certain quality of life that they have determined to be acceptable. AI CDS is best suited to make predictions based on an objective outcome. That outcome may or may not be the outcome that the patient seeks to achieve. In addition, patients may desire the same outcome as the AI CDS is programmed to predict but may find some means of achieving that outcome unacceptable. For instance, a patient who is undergoing cardiac surgery may desire to extend their survival but find the transfusion of blood products unacceptable to achieve that outcome. While the patient may value the same outcome that the AI CDS is programmed to achieve, extended survival, the patient may refuse the means of achieving that outcome, cardiac surgery, if there is a high risk that the operation will require the transfusion of blood products.

Introducing AI CDS into the physician-patient relationship and the process of shared decision-making does not necessarily exclude patient values from being taken into account when making a final decision. There is nothing about the system's prediction in and of itself that binds the patient to share that desired outcome or the means of achieving it. However, depending on how these systems are integrated into the healthcare system, it may well be that extrinsic forces such as payers or health systems will limit the extent to which patient values will be accommodated. The systems themselves might be encoded to prioritize nonclinical outcomes, such as cost savings or efficiency, the achievement of which ignores patient preferences and values (see also "Conflicts of Interest" section in this chapter). This could affect patients in two different ways. Patients could be refused either their value-driven desired outcome or their desired means of achieving that outcome. In addition, patients may have no other choice but to accept an outcome or means of achieving the outcome that is not in keeping with their values and preferences. Payers may choose to only reimburse that which the system predicts, regardless of patient preferences. Health systems may expect physicians to achieve patient compliance with system predictions in order to take advantage of its cost-effectiveness or reduce the risk of medical-legal liability.

In addition to taking into account patient values when deciding on a clinical course of action, it is also important to consider patient values in allowing patients to decide whether or not it is acceptable for their physician to collaborate with an AI system in making medical decisions or predictions.[11] This is especially important when AI systems are embedded in the EMR to make administrative predictions about patients that do not directly impact their immediate health care. Take for instance an AI system that predicts that a patient will likely die within six months. The goal of the prediction may be to identify patients at highest risk of dying so as to assist them with advance care planning and match them to palliative care resources. A second and related goal of the system may be to reduce hospital mortality rates, a commonly reported marker of a hospital's quality (higher quality hospitals are assumed to have lower mortality rates). The system is automated to make the prediction for every patient admitted to the hospital.

While such a system raises multiple ethical concerns, relevant to this discussion is that some patients may not want to know if they are predicted to live less than six months. In addition, some patients may not want such a prediction to be made at all, since once the prediction is made, it becomes part of the patient's medical record and has the potential to influence decisions made about the patient's medical care. Such a prediction could cause clinicians to view some treatments as futile since the patient is going to die within six months. In order for patients to decide whether or not AI systems are acceptable, they should be informed about the system and the prediction that it is making. Respecting patient autonomy has always prioritized the patient's right to refuse anything that they find unacceptable. There may be instances in which a patient's refusal of some part of the treatment would necessarily mean that the treatment itself could not be delivered. For instance, a patient might accept an intravenous medication but refuse to allow an intravenous catheter to be inserted in order for the medication to be infused. As AI systems are being integrated into health systems, it is important to think about how to accommodate a patient's refusal. This of course demands that patients be informed about the use of these systems.

THE PATIENT AS HEALTH INFORMATION DONOR

The discussion up to this point has focused on the importance of informing the patient qua patient in order to establish and maintain trust in the therapeutic relationship between patient and physician. However, when AI CDS is introduced into the physician-patient relationship, the patient also takes on other roles besides the patient role. In order to receive a prediction from AI CDS, the patient's personal health information must be transferred to the system. This transfer occurs not just when the patient receives the prediction, but is ongoing in order to determine the accuracy of the system's prediction. For instance, a system that predicts six-month mortality may predict that 17 percent of patients admitted to a particular hospital ward will die within six months. In order to test the accuracy of the system's predictions, the patients' future health care data will need to be transferred to the system in order to determine whether the 17 percent of patients predicted to die within six months actually do die as the system predicted.

The transfer of medical information to the system has important ethical implications, beyond considerations of privacy, which is discussed later in this chapter. Transferring a patient's health information to an AI system very much resembles a donation in several ethically relevant ways. First, the patient's personal health information is passed from the patient to the system without the expectation that the information will be returned or even the ability to return it if so desired. The patient's information is transferred in one direction. Also, donations are inherently relational. In this case the patient relates to the system both by donating their health information to it and in receiving a prediction back from the system. In both instances, the transfer involves the patient's physician. The physician, as noted previously, has the primary direct relationship with the patient and the system, with the system having an indirect relationship with the patient through the physician. This is true both in transferring information from the physician to the system and in the physician's collaborating with the system in order to make a decision that will ultimately affect the patient. Third, and perhaps most important, donations have value.

What is it about a patient's health information that gives it value? If human genetic material, a patient's genotype, is the repository that contains unique information about what a person has the potential to be, a patient's personal health information is the expression of that genetic material, the person's phenotype, who the person is. In the case of a patient's EMR, it is their digital phenotype. A patient's EMR contains their personal health information, which is essentially the lived experience of how the patient's unique genetic material expressed itself. Each person's EMR contains an account of who they are, in many instances today, from birth to death. It tells the story of how a person's genes were expressed, interacted with the environment, and shaped the course of their lives. For most people, it is the structure around which their personal narratives are constructed. For many, some of the greatest joys of their lives, such as childbirth, adoption, surrogacy, and relationships, are contained in that record, as well as their greatest sorrows, such as disease, serious illness, physical and emotional trauma, injury, suffering, and death. In many instances, the patient may have shared the details of these events only with their physician, who inscribed them into the patient's EMR. While each of these individual facts about a person may be accessible, it is only in their medical record that one can find the complete story. It is in being an individual, personal account of the patient that the EMR has value.

Donors Must Be Informed

Classifying the patient as a donor of their personal health information implies certain ethical obligations beyond that which is due to them as a patient. For instance, when Sam is admitted to the hospital in order to remove a kidney that he is donating to his sister, Sam is not only a surgical patient but he is also a donor, and there are ethical considerations relevant to Sam as a donor that are distinct from the ethical considerations relevant to Sam as a patient. Similarly, when a patient is a donor of their health information to an AI system, several important ethical considerations are relevant to them, most notably that donors are necessarily informed and that their donation is voluntary.

The obligation that donors be informed prior to making a donation is another argument in favor of explaining to patients not only that their physician will collaborate with an AI system, but also that the cooperation requires that the patient donate their data to the system.[12] The risks, benefits, and alternatives of each are distinct. While the risks and benefits of using AI CDS to make clinical decisions are largely the project of this work, the specific risks and benefits of health information donation itself bear consideration also. Chief among the risks to the patient of donating their health information is a breach of confidentiality, which would risk harming the patient directly and also harming the physician-patient relationship.

The Risks of Donating Health Information

When taking the Hippocratic Oath every physician swears, "Whatever I see or hear, professionally or privately, which ought not to be divulged, I will keep secret and tell no one."[13] This profession has been interpreted to mean that physicians are bound to hold information which they know about the patient in confidence. When the relationship between the patient and the physician was verbal, this meant not repeating what the patient told the physician. When medical notes were written in charts, in addition to verbal confidence, the obligation extended to restricting access to the patient's written chart. Only those involved in the care of the patient could rightfully have access. Currently the patient's health information is all electronically stored. In some jurisdictions, laws may designate patients as the owners of their health information. That information may include a physician's summary of their conversation with the patient, laboratory and imaging results, genetic testing, and increasingly, the predictions of AI CDS. The point here is that although patients' health information has become less able to be directly controlled by the physician, the obligation to keep confidence has not changed; it has just required different means of fulfilling the oath. From auricular to written to digital, the patient with whom the physician has a relationship and to whom the physician owes the obligation of their oath can rightfully expect that whatever the physician sees or hears (ἴδω ἢ ἀκούσω) they will keep secret (σιγήσομαι). Breaking this oath not only places moral blame on the physician but also, inevitably, erodes the patient's trust in the relationship with their physician.

If in the auricular age the oath was interpreted to mean that physicians should not verbally repeat information about the patient, and in the written age it was interpreted to mean that only those caring for the patient could read the patient's written medical record, what is the application of the oath in the age of AI CDS? The most common interpretation is that health information that has been donated by the patient should be de-identified if it is used for purposes other than the direct care of the patient, which may imply that health systems are either sharing the health information that patients have donated with AI system developers in order to train and validate new AI systems or selling it to them.[14]

Besides compliance with legal requirements, such as the Health Insurance Portability and Accountability Act (HIPAA), the reason for de-identifying patient data is to maintain confidentiality and protect patient privacy when data is transferred out of the protected electronic health record. The ease with which patient health information is reidentified raises legitimate questions about how "de-identified" patient health information actually is.[15] Sharing or selling a patient's donated health information raises ethical concerns besides a breach in confidentiality that is addressed later in this chapter. It seems that the greatest and most direct risk of harm to a patient donating their health information, as it pertains to confidentiality in the age of AI CDS, is the possibility that clinical predictions made by one system could influence the prediction made by another system so as to limit the options available to the patient.[16] Even if there is no direct sharing of predictions across systems, one system's prediction could profile the patient in such a way as to also impact other predictions. The issue here is twofold. First, even if the patient was informed and consented to their physician's collaborating with an AI system to make a clinical decision, their consent was decision specific and would not necessarily give the system permission to share their health information with other systems. The second concern is that these are only predictions and not the actual reality of the patient. The risk of harm when a system shares a patient's health information across systems, or profiles the patient, is that predictions could become necessarily self-fulfilling even though that was not the inevitable outcome.[17]

Take for instance the AI CDS that predicts six-month mortality. If a patient with acute liver failure presented to the hospital, without appropriate treatment that patient's survival would be less than six months. The mortality prediction made by the system could then influence other predictions based on the assumption that the patient will die within six months. Even more concerning is that the mortality predicting system might share its prediction with a payer's system and thus indirectly limit options available to the patient by denying reimbursement under the assumption that the patient will die within six months. Similarly, a system that predicts which patients are at risk of developing atrial fibrillation may share that prediction with an AI CDS that predicts who should receive complex cardiac surgery. Even though the patient does not actually have atrial fibrillation and is only predicted to be at risk for developing it in the future, such a prediction runs the risk of biasing an AI CDS for cardiac surgery. Thus, the patient may not be offered a potentially beneficial surgery based on a prediction. Furthermore, because the cardiac surgery prediction is not explainable, it will not be clear why the prediction was made so that it can be corrected.

The Benefits of Donating Health Information

The presumed benefit of donating one's digital phenotype to the AI CDS is to receive a medical prediction that is believed to be more accurate, precise, and personalized than a physician acting alone could make. Reducing medical uncertainty would thus be beneficial to the patient. While this presumption must be tested in order to be accepted, it does seem reasonable to assume that some systems will be markedly beneficial to patients. Altruistically, patients may perceive a benefit from contributing their health information to a system that relies on information to make better future predictions. Patients may believe that by donating they will contribute positively to future iterations of the system that will be even more beneficial to patients.

The Alternative to Health Information Donation

While it may be clear that some AI CDS will be beneficial to patients and to society at large, what is less clear is what alternatives there will be to using AI CDS to make clinical decisions once these systems are embedded in the health system. It is widely accepted that patients have the right to informed refusal in medicine, even if what they are being offered is proven to be more beneficial than the alternative.[18] This is again where patient values come into the consideration of what is beneficial and what poses an unacceptable risk of harm. Patients who refuse to donate their health information and thus refuse to allow their physicians to collaborate with the system may find that there is no clear alternative.[19] If the alternative of having a physician make medical decisions without collaboration with the AI system is available, the patient may find that their physician's decision-making skills have dulled due to overreliance on the AI system. This poses an additional harm that patients ought to be informed of as it may factor into their decision-making about accepting or refusing to donate their health information.

Donations Must Be Voluntary

This leads to consideration of the other key ethical obligation owed to donors, namely that the donation be voluntary. There are two important factors that influence the voluntariness of a patient's donation of their health information to an AI CDS system. First, a patient's health information donation is the sine qua non for allowing collaboration between the system and the physician and receiving a presumably more beneficial decision than the physician alone could provide. This sets up a quid pro quo that may significantly limit the extent to which the patient's donation is voluntary. Second, as discussed earlier, how voluntary can a donation be if either a viable alternative doesn't exist or the viable alternative is unduly risky? In order to make the patient's donation as voluntary as possible, they must first be informed of the donation, given an accurate account of the risks and benefits, and be presented with a reliably beneficial alternative.

The stories of Henrietta Lacks and John Moore serve as a guide when considering how important it is that donations of any sort from a patient be voluntary and informed. This is true no matter how inconsequential those procuring the donation may believe the donation to be, even if they don't believe that which is being procured is a donation at all. In the case of Ms. Lacks, cancerous cells from her cervix, and in the case of Mr. Moore, cells from his spleen, were taken from the operating room to the laboratory by physicians for medical research. In neither instance was the donation informed or voluntary.[20]

THE PATIENT AS A LEARNING SUBJECT

Many AI systems that are used to make clinical predictions are continuously learning from the predictions they make. As noted previously, after the patient has donated their health information to the system, the system has ongoing access to the patient's EMR so that it can learn whether or not its prediction was accurate. This allows the system to improve its accuracy with time. The systems are referred to as machine learning because they are continuously learning. In many health systems there are medical and nursing students, resident physicians, and fellows, all of whom to one degree or another are also using the patient as a subject for their learning. It is well established that when a patient is also a learning subject for human learning in the context of medical education, their participation should be both informed and voluntary.[21] Similarly, when a patient donates their health information to an AI system and receives a clinical prediction from it, they are also learning subjects for the system, and their participation should be both informed and voluntary.

Just as with the role of the patient as patient and the role of patient as donor, when the patient is in the role of a learning subject distinct ethical obligations apply. While there is clear overlap in the ethical obligations owed to these roles, particularly the roles of donor and learning subject, it is important to consider that when receiving a clinical decision from an AI CDS system, the patient is not able to decide which role they wish to participate in and which they do not. The three roles are necessarily and inextricably commingled. In consenting to their physician's collaboration with an AI system, the patient is also consenting to be a data donor and a learning subject. In contrast, a patient who does not consent to the collaboration between physician and machine can choose to be a patient, to be a patient who is also a donor, to be a patient who is also a learning subject, or to be a patient who is both a learning subject and a donor. The importance of this point is to stress that using AI CDS to make a patient's medical decisions should be both informed and voluntary.

THE PATIENT AS A RESEARCH SUBJECT

There may be instances in which the clinical benefit of AI CDS is being studied within the context of a research trial. In such a case, the role of the patient as a donor, a learning subject, and a research subject would necessarily be informed and voluntary because there are well-established rules governing participants in research studies. It is notable that currently the only role that would require that patient participation with AI CDS be informed and voluntary is in the context of a research study. While there are clear ethical obligations that govern each of the roles that the patient plays relative to the AI system, it is the role as research subject that is most tightly regulated and has the greatest oversight, largely because of unethical treatment of research subjects in the past.[22] This observation should be taken seriously as AI CDS is deployed more broadly. If patients are not informed as patients, donors, and learning subjects about the use of AI CDS in making their medical decisions, and if their participation is not voluntary, then it is very likely that in hindsight it will be clear that the same unethical practices that were previously taken for granted as standard procedure in medical research are being played out once again with the introduction of AI CDS. Furthermore, the erosion of trust in medicine and health-care systems that resulted from previous unethical behavior is still impacting society today. Any further loss of trust will be even more difficult, if not impossible, to repair.

THE PHYSICIAN'S FIDUCIARY RESPONSIBILITY: PRIORITIZING THE INTERESTS OF PATIENTS

Patients rightfully expect their physicians to prioritize their well-being. Not only is it the fiduciary responsibility of the physician to prioritize the patient's well-being over other competing priorities, but this also allows patients to place trust in their physicians. When a patient consults with a physician, they expect that the physician's decision is free of motives beyond the patient's best interest, as defined by the patient, and that the physician's decision is objective and unbiased. Fulfillment of these rightful expectations leads to further trust and growth in the relationship. When patients believe that their physician has a competing interest beyond their well-being, or if they believe that the physician is biased in the decision they are making, they will inevitably come to distrust their physician and the therapeutic relationship will be ruptured.

Conflicts of Interest Involving Physician Collaboration with AI CDS

How might physicians either not act, or be perceived as not acting, in a patient's best interest; how might they be biased in their decision-making; and how does this apply to the introduction of AI CDS into the relationship? At the most basic level, physicians may be perceived as not acting in a patient's best interest if they have an additional interest to which they either give equal priority or if they put that interest ahead of the patient's. Conflicts of interest such as this often involve a financial secondary interest, but not always. Secondary interests can also be valuable in other ways such as organizational advancement, status, or prestige.

Take for instance a medical oncologist who treats breast cancer using chemotherapy and who is also a paid consultant for a drug company that makes a drug to treat breast cancer. Patients may question whether the physician is prescribing a particular drug for them because the physician believes it is the most beneficial or because the physician receives money from the drug company. Studies have shown that even gifts from drug companies with little monetary value can influence physicians in favor of prescribing that company's drug.[23] A perceived conflict of interest may also exist when the health system a physician belongs to expects them to only refer to other physicians within that system, regardless of their level of expertise. Physicians who comply with the system's expectations may be more likely to be rewarded with advancement and administrative jobs than those physicians who do not comply. Patients may come to lose trust in their physicians if they believe that the reason they are being referred to another physician is that the physician is part of the same system, rather than believing that they are being referred because the physician is the best person to address their medical needs.

Patients may perceive that there is a conflict of interest when their physician consults with AI CDS in at least two different ways. First, patients may question whether the decision made by the AI CDS prioritizes their well-being over other interests such as cost savings, efficiency, or improving a health system's outcome metrics and reputation. Patients also may perceive that there is a conflict if they believe there is extrinsic pressure on their physician to accept the system's decision and to encourage the patient to do likewise, regardless of whether or not the physician believes that the decision is actually most beneficial for the patient.

Because the weight given to the various interests that have been programmed into the system are often opaque, patients may not have a basis for trusting that the system prioritizes their own interest over other interests. Since the physician has the primary relationship with the system and cooperates with the system, the patient's mistrust may be transferred to the physician. The patient may believe that the physician and the health system are colluding in order to save money, increase efficiency, or improve the health system's reputation at the patient's expense.

Take for instance the previously described AI CDS that predicts six-month mortality. The health system instituting the AI CDS may have as its goal reducing the number of people who die in the hospital so as to improve the health system's reputation, since lower hospital mortality is associated with better quality of patient care. The AI CDS may be explained to patients as a way to ensure that their values are respected and that they are matched with appropriate support services at the end of their lives. Patients may not know which of these competing interests is prioritized, which can lead to distrust. Health systems and their physicians can be proactive by being transparent about the different interests and truthful in assuring patients that their benefit is prioritized over the health system's reputation. Physicians and health systems can certainly have other interests besides the patient's well-being, such as cost savings and efficiency. However, the interest of the patient must always be prioritized over the other interests. If competing interests are hidden from patients, or if the patient's benefit is not truly prioritized, then patients will lose trust in the health system and in their physicians.

A patient may also come to distrust their physician if they believe that the physician has an interest in having the patient accept the AI system's decision even if the physician does not think it is the most beneficial option for them. Examples are if physicians are reimbursed at a higher rate because their patients accept the decisions made by the system, or if physicians are more likely to be promoted within their medical group if their patients follow the decisions made by the system. Even if there is no direct benefit to the physician in terms of financial gain or advancement, a physician may nonetheless feel as though patient acceptance of the AI system's decision will protect them in case of a medical-legal claim. This may lead a physician to believe they need to "sell" the AI CDS decision to their patient without regard for whether the physician believes that the decision made by the system is actually most beneficial for the patient. Once again, this is liable to lead to distrust and erosion of the patient-physician relationship.

AVOIDING BIAS WHEN PHYSICIANS COLLABORATE WITH AI CDS

Patients trust that when a physician makes a medical decision about them, not only does the physician prioritize their best interest, but the decision is also objective and unbiased. Physicians often speak of "one standard of care for all" as an application of Beauchamp and Childress's principle of justice, meaning that diagnoses will be made objectively and the accepted treatment for a given illness will be offered to every patient in the same way and without bias.[24] When physicians fail to fulfill this obligation to their patients, patients lose trust in their physicians. While AI CDS holds great promise for reducing human bias in medical decision-making, without careful attention to the ways in which bias can be encoded and perpetuated by these systems, they will fail in their promise of objectivity and will come to be distrusted by patients and their physicians alike.[25]

There are two important ways in which AI systems could render biased decisions. First, if the wrong algorithmic assumption is made, the decision may be biased. Second, AI CDS may be biased if the dataset on which the system is trained contains data that are unrepresentative of the patients for whom the system is making predictions. Ziad Obermeyer and colleagues, in their foundational paper, showed how bias could result when an algorithm incorrectly assumed that previous health-care expenses correlated with a patient's severity of illness.[26] The system described by Obermeyer and colleagues was used to provide patients deemed to have the most serious illnesses with additional support upon discharge in order to prevent them from being readmitted to the hospital. The system assumed past health-care expenses correlated with severity of illness and thus provided those patients with the highest past health-care expenses with the most additional resources upon discharge. The outcome was that many White patients with insurance had the highest previous health-care expenses and were provided with the most assistance at discharge. In contrast, many Black patients who were uninsured or underinsured had lower previous health-care expenses and were given fewer resources at discharge, despite having overall more serious illness than did White patients. The wrong algorithmic assumption led to marked bias in the distribution of resources. This example highlights the importance of distinguishing an observed correlation between two things from a factual occurrence in which one thing actually causes the other thing to happen.

An unrepresentative dataset can also lead to a biased decision. How might a dataset be unrepresentative of patients for whom it is being used to make decisions? One way is that characteristics of the data on which the system is trained, such as race, gender, age, and disability status, may not reflect the characteristics of the patients for whom the system is making predictions. A system trained on a dataset consisting of mostly White patients may perform poorly when making decisions for mostly Black patients.[27] Datasets can be unrepresentative in several ways. First, the data may not have been sampled correctly, or there may have been an error in data measurement.[28] Unequal access to health care and physician bias can also result in skewed data. For instance, if Black patients with cancer are less likely than similar White patients to be offered curative surgery because surgeons perceive that Black patients are at higher risk for death or complications than White patients, the dataset will reflect a higher cancer-associated mortality rate for Black patients who did not receive a curative treatment than for White patients who did receive a curative treatment.[29]

Several steps can be taken to assure that the decisions made by AI CDS are objective and unbiased and do not lead to distrust. First, algorithmic assumptions must be tested to make sure that they won't lead to biased and inequitable outcomes. Second, historic data need to be representative and free of previously biased decisions made by clinicians. This will require much more effort because it will involve analyzing larger and more complex datasets. One ethically supported solution is to set limits on the applicability of datasets.[30] For instance, if a particular AI CDS system has been trained on data from mostly men in the United States, then it would be limited to providing clinical decisions about men in the United States. Setting such limits would avoid the risk that an invalid prediction would be made for a patient based on an unrepresentative dataset, and it might also spur developers to invest in analyzing larger and more complex datasets with wider applicability.

The patient's health information is given to the system in order for the patient to receive the desired prediction. In fact, the patient's health information may also be sold by the health system to developers so that the systems have adequate data on which to train and validate what they are programming.[31] Although the data are typically de-identified—meaning, at a minimum, that the patient's name, date of birth, medical record number, and social security number have been removed—the data nonetheless are those of the patient.

THE PHYSICIAN'S COMMITMENT TO NONABANDONMENT

Professing to not abandon their patients, along with professing to relieve their suffering, is among the most important things that physicians can do.[32] These commitments are essential to establishing and nurturing trust between patients and physicians. The commitment to nonabandonment implies that the physician will engage with the patient and personally care for them until the patient no longer either desires or requires the physician's care. Given the importance of this commitment to cultivating trust, how might it be interpreted when introducing AI CDS into the patient-physician relationship?

Some of the ways in which physicians might abandon their patients have been mentioned previously. Notably, they all involve the physician abandoning the essential function of the physician, which is medical decision-making.[33] For example, a physician may seem to abandon their patient when automation bias leads them to automatically accept the AI system's decision without question. The physician may sincerely believe that the decision-making ability of AI CDS is superior to the physician's decision-making ability. This may be because of the amount of data the system was trained on, its method of validation, and its accuracy in making clinical decisions. Physicians may seem to be abandoning their patients if there is external pressure for them to ensure that patients comply with the decisions of AI CDS. However, in the instance of automation bias, the physician may truly believe that they are doing the most good for and least harm to the patient by following the decision of the AI system without question. In the instance of external pressure, it may well be that the physician is experiencing moral distress. The physician knows that the right thing to do is to exercise their own judgment, but they are unable to do so because of forces beyond their control. Physicians can thus either comply, or they can quit, the latter of which could lead to patients feeling even greater abandonment.

It seems that the most clear instance of a physician abandoning their patient once AI CDS is introduced into the relationship would occur when physicians relinquish their judgment to the AI system, so called automation complacency.[34] This sort of abandonment could occur in multiple ways along a continuum. At one end of the continuum the physician is overwhelmed by multiple tasks and accepts the AI CDS decision outright. The physician justifies their decision because the system has been accurate in the past, and accepting the system's decision without evaluating the clinical data and making an independent decision will save time and allow the physician to complete other tasks. At the other end of the spectrum, the physician becomes apathetic and nihilistic and stops caring about their role. They may believe that they have been replaced; that they are a cog in the system's machinery; and that regardless of what decision they make, they must implement the decision of the AI CDS system. What this physician forgets is that regardless of the complexity of their relationship with AI CDS, their primary duty is to the patient. Moreover, by relinquishing their primary role, medical decision-making, to AI CDS without regard for the patient, the physician in fact abandons both their primary role and their primary relationship, that with the patient.

As AI systems are deployed, every effort should be made to prevent both automation bias and external pressure on the physician and patient to accept the AI system's decisions without question. Doing so will foster a patient's trust not only that their physician has not abandoned them, but also that their physician is not prioritizing other interests over the patient's own benefit. Physicians must also be steadfast in their commitment to not abandon patients, even if that means exercising their essential role of medical decision maker when that role may seem devalued. In abandoning their role as medical decision makers they would also be abandoning their patients, their oath, and their personal identity as physicians.

TRUTH AND TRANSPARENCY

In every human relationship, trust is fostered by truthfulness. For physicians, truthfulness entails being honest with patients, being transparent, and not lying. Implicit in the previous account of ways in which physicians establish and maintain trust with their patients has been the central role of truthfulness and transparency. However, certain applications of honesty and transparency to AI CDS deserve highlighting. This section hones in on the importance of disclosing to patients when AI CDS is used to make medical decisions, being honest with patients if there are conflicts of interest, being transparent about the selling of donated medical information, and being truthful when the system's decisions are not accurate for a certain group of people either because they are underrepresented in the dataset or because the data show a pattern of inequity.

Not only is informing patients one of the primary ethical obligations owed by the physician to the patient, a logical extension of the physician's essential role of decision-making, but it also fulfills the physician's obligation to be truthful. The virtue of honesty is manifest in the patient-physician relationship by being honest and transparent with the patient. It is also generally recognized that patients can decide for themselves what "truths," in the form of their medical information, are important to them and what "truths" are not important for the purpose of engaging in shared decision-making. The point here is less to establish criteria for disclosure of truths than to highlight that part of respecting patients' values is allowing them to play a role in deciding what information they value when making a decision about their health. To dc facto determine that patients do not want or need to be informed regarding the collaboration between their physician and AI CDS in making their medical decisions fails in the duty to inform patients, and such a claim is also dishonest and will inevitably lead to distrust.

As has been noted, when physicians collaborate with AI CDS, there may be conflicts of interest in which other outcomes are either of equal or possibly greater priority than the benefit of the patient. If the physician is aware of such conflicts, they must be honest about them and disclose such conflicts to the patient. Although most of the conflicts of interest will either be as opaque to the physician as they are to the patient or beyond the control of the physician, it is nonetheless the physician who has the primary relationship with the patient, it is the physician who has the greatest ethical obligation to the patient, and it is the physician who will lose the patient's trust and the therapeutic alliance if conflicts of interest are not disclosed to the patient.

Although it is not necessarily a conflict of interest, patients may feel duped or used when health systems sell their donated health information to AI developers without either informing the patients or getting their expressed consent to having their data sold, even if the information is de-identified.[35] As has been argued, a patient's health information is their digital phenotype and as such has personal value. Selling that information without the informed consent of the patient, particularly after it has been donated, is an act of dishonesty and will lead to distrust.[36]

Because there are datasets that are either unrepresentative of certain groups of patients or in which the data are biased because of systemic inequities in health care, AI systems trained on such datasets may not provide accurate predictions to some groups of people. As such, when data are known to be unrepresentative or biased it would be dishonest to withhold such information. Doing so may not only be dishonest, but may also be directly harmful to those groups of patients.

ACCOUNTABILITY FOR THE PHYSICIAN-AI COLLABORATION

Trust is also built in the physician-patient relationship when patients know that physicians are accountable for the medical decisions they make. The range of accountability can encompass personal moral responsibility, admission of error, and personal apology at one end, and litigation, medical board penalty, and loss of professional employment at the other extreme. Such consequences typically occur when a physician has made a medical error that resulted in some level of harm. If the error was within the standard of care and the physician was transparent and offered an apology, the likely outcome would be less severe than if the error was a substantial deviation from the standard of care, the physician tried to cover up the error, and/or the physician shrugged off their personal responsibility for harming the patient.

Just as patient trust is engendered by physician accountability, so too will accountability for the decisions that stem from the physician-AI system collaboration foster trust.[37] What is less clear is how accountability for harm caused by collaborations between the AI system and the physician will come to be understood. Because the collaboration is between two entities, one human and one machine, it seems unlikely that the physician would always be entirely accountable for errors and harm even though the physician does have the primary relationship with the patient and owes to the patient a defined ethical obligation.

When might a physician be held primarily accountable? One legal perspective is that once AI systems become the standard of care, if a physician makes a decision contrary to that made by the AI system, and the patient is harmed by that decision, then the physician would likely be held accountable.[38] This reflects the current practice in which if the physician acted contrary to the standard of care, which led to the patient being harmed, the physician would be liable.[39]

How might accountability be assigned if either the AI system had not been established as the standard of care or it had become the standard of care and the patient was harmed because the physician enacted the system's decision? As discussed previously, it seems unjustified that the physician would always accept full responsibility for harm if the harm was due to an error in the system. An example is if the system was trained on a dataset that was unrepresentative of the patient for whom the decision was being made and the prediction made by the system was not accurate for that patient. Unless limits were set on using the AI CDS to make decisions for certain patients such as the one in question, the programmer and developer would seem to be more at fault than the physician. The conclusion might be different if limits were set but were ignored by the physician.

Although the locus of collaboration is between the physician and the AI system, there are multiple other stakeholders involved in conceptualizing, developing, programming, validating, adopting, and deploying the system. Because of the role each of these stakeholders plays, often behind the scenes, in bringing the AI system into collaboration with the physician, it seems that there is rightfully diffused responsibility and accountability among multiple stakeholders.[40] One proposed solution to the issue of accountability shared among multiple stakeholders is a participatory approach in which the stakeholders themselves decide how to assign accountability for errors and harms.[41] Based on the success of a participatory approach in AI system design, as demonstrated by WeBuildAI, there is every reason to believe that application of such a model to assigning accountability would be just as effective.[42] Of particular importance is including the patient as the primary stakeholder in these discussions. Too often in the history of medicine has the patient been assigned the role of passive recipient of medical truth. A much more ethically tenable model, particularly with the introduction of AI CDS, is to include the patient not only in the process of shared decision-making about a discrete medical decision, but also more broadly in the process by which medical decisions are being made and the means by which accountability for those decisions is being assigned.

CONCLUSION

As AI systems have been deployed in various sectors, there has been a concerted effort to both define and engender trust in these systems. As AI is introduced into the physician-patient relationship, it is important to reflect on the ways in which patients come to trust, and continue to trust, their physicians. Application of these actions to the collaboration between physicians and AI systems—informing, making participation voluntary, maintaining confidentiality, prioritizing the patient's well-being, avoiding conflicts of interest and bias, not abandoning the patient, and being truthful and accountable—is not only the obligation of the physician, but it is also how patients will come to view favorably the introduction of AI systems into the patient-physician relationship.

CASES

Case Study 3.1: Patient Entrustment of AI-Enabled Decision Support

Mr. Hickory is a seventy-three-year-old retired Alaskan steelworker with a lung mass. Two weeks ago he developed fever and a cough producing green-tinged sputum, prompting his visit to an urgent care center. A chest radiograph confirmed the suspected diagnosis of pneumonia and also identified a 2 cm mass in the central portion of his right lung. His primary care provider advised that he seek care at a large medical center.

Early one morning, Mr. Hickory boards a plane bound for Seattle. He doesn't much like planes but takes some comfort in the flight attendant's warm smile and a glance over her shoulder at the cockpit, where he sees the pilot and copilot carefully reviewing their preflight checklist. Mr. Hickory understands—with some consternation—that most of his flight will be on autopilot, but appreciates that two highly trained human beings will be vigilant in overseeing the entire flight and guiding the plane with their own hands during takeoff and landing.

That afternoon, Mr. Hickory checks into a clinic for his initial consultation. Unlike most of the patients standing in line, he reads all ten pages of the waiver he's asked to sign. Most of the waiver appears to be lawyer-derived boilerplate detailing the hospital's process for disclaiming liability for many aspects of the care that will be provided. He grimaces but remembers that his lawyer daughter-in-law says such waivers provide necessary protections for the hospital. Then there is a section about how data collected for his medical care might be used for research. This sounds like someone will be experimenting on him. Mr. Hickory is ready to rip the waiver lengthwise, but his curiosity wins out, and he continues reading, learning that no experimental interventions will be performed without his express written consent. The final pages bring another surprise: his physicians may consult with predictive analytic tools that use AI algorithms. There isn't much detail about how the algorithms work or how they will be used. This time, his desire to rip the waiver lengthwise is overcome by curiosity. He begrudgingly signs the waiver. After he spends another twenty minutes in the waiting room, a nurse checks his vital signs. She inquires, "Does your blood pressure usually run high?" He frowns. "No." The nurse leads Mr. Hickory to the exam room where the doctor will see him.

Mr. Hickory is still frowning when his physician knocks on the door and enters the exam room with a smile that fails to penetrate Mr. Hickory's angst. Before discussing the lung mass, she warmly invites Mr. Hickory's description of his present understanding of the situation. He has read several articles about lung masses and feels fairly well informed about them, and wishes instead to discuss his permission for his physicians to consult with AI-enabled predictive analytic tools. "How do they work?", he demands. His physician replies that the AI algorithms use patient data and complex mathematical formulas to make predictions about how a patient will respond to a treatment. He presses further, "How do you know it's using the right patient data and making an accurate prediction?" She admits that she and other physicians using the AI systems don't know whether the system is using the right data or making an accurate prediction; they just know that the systems perform well overall. He responds by saying, "You know they perform well based on what—the AI science experiments you're doing on me and other patients? Well, I'm not a guinea pig and I'm going to find a doctor who knows what to do without letting an AI robot do their thinking for them, without showing its work."

QUESTIONS

The hospital and the physician attempted to be transparent and informative about their use of AI in providing care. Were their attempts sufficient? Why or why not?

What are the ethical implications and principles involved in allowing patients to opt out of AI CDS use?

Case Study 3.2: Protecting Patient Privacy while Using AI

Ms. Edwards is a forty-seven-year-old middle school teacher whose family tree has been riddled with both rare and common cancers. At the advice of her primary care provider, Ms. Edwards seeks consultation with Dr. Gentry, a genetic counselor who uses medical and family histories and genetic testing to provide knowledge and wisdom to patients regarding personal and familial risk for diseases that have genetic causes. To provide that knowledge and wisdom, Dr. Gentry depends upon calculations of probabilities that a given gene sequence will manifest as a disease and probabilities that two given gene sequences, when combined in forming a new life, will generate progeny gene sequences that will manifest as a disease. Calculating these probabilities is aided substantially by leveraging AI algorithms. Dr. Gentry participates in multicenter research in which AI uses genetic data to predict disease manifestations in individuals and family members.

Preserving patient privacy is a high priority in all health-care applications. For gene-related health-care applications, it's an even higher priority, because genetic information has important implications for establishing paternity, choosing whether to proceed with live birth of a fetus, and discrimination in job offers and insurance coverage. In addition to these issues, each individual has the right to remain unaware of their genetic makeup, and when a family member undergoes genetic testing, there is risk for inadvertent or intentional sharing of the family member's genetic information with the individual who chooses not to know about it.

Cognizant and respectful of patient privacy in genetic testing, Dr. Gentry is careful to collaborate only with other investigators who, like him, use only the personally identifying information (i.e., the type of information that, in combination with other information from voter registration databases or social media, could be used to identify a single person) that is necessary for the research being performed, and who also use password protections and encryption software and communication protocols that minimize the risk of disclosing information that adversaries could use to identify patients.

Despite best efforts by the investigators, an adversary silently inserts computer software into the chain of computers that perform multi-institutional analyses of patients' genetic data. Investigation of the data upstream and downstream of the spyware insertion suggests that the data were not altered, but it is possible that the data were collected and saved by the adversary.

QUESTIONS

Are Dr. Gentry and his collaborators obligated to notify all patients who were potentially affected that their information may have been disclosed, or to pay a ransom demanded by the adversary for returning the information?

What are the potential effects on family members of patients whose genetic data were disclosed, particularly if they choose to remain unaware of their own genetic data?

Case Study 3.3: Human-Machine Teaming and Discordance in Shared Medical Decision-Making

Patients with "resectable" pancreatic cancer can be treated with a combination of surgery, systemic chemotherapy, and radiation therapy. There is debate about whether surgery should be performed first or after several doses of chemotherapy. There is also debate about whether radiation therapy confers any benefit. Several chemotherapy drugs are effective against pancreatic cancer. One of the most common regimens is three months of combination chemotherapy (agents U, F, and O) along with radiation, followed by surgery and three additional months of chemotherapy (agents U, F, and O).

A seventy-two-year-old, healthy male patient is diagnosed with a "resectable" pancreatic cancer. The patient sees Dr. King, a renowned pancreatic cancer specialist at Premier Cancer Center. As is his custom, Dr. King presents his patient's case to a group of his colleagues for their treatment recommendations. The surgeon recommends immediate surgery, followed by chemotherapy for six months (agents U, F, and O) but no radiation therapy. Three medical oncologists make three different recommendations for combinations of chemotherapy (agents U, F, and O; agents F and G; agents U and F) along with radiation therapy for three months, followed by surgery, followed by three more months of chemotherapy. The medical oncologists recommending only two chemotherapy agents think that all three agents will cause too much harm to the patient from chemotherapy side effects. The radiation specialist recommends chemotherapy (agent G) for six months along with radiation, followed by surgery and additional radiation.

Premier Cancer Center has invested in an AI CDS system that predicts treatment plans with the goal of best long-term survival. The CEO of Premier Cancer Center is on the board of directors of OncoAI, the AI CDS system developer. She receives approximately $800,000 per year in stock options for her service on the board. She is under a tremendous amount of pressure to decrease the cost of cancer treatment. OncoAI openly claims that the system both reduces cost and extends survival over traditional treatment planning performed by physicians. The system has never been subject to a clinical trial, and so there is no high-level evidence to guide the decision-making process.

The system makes a treatment prediction for Dr. King's patient that is different from the treatments recommended by the specialists. The system recommends systemic chemotherapy with agents F and O for two months, followed by two months of radiation combined with agent G, then surgery, followed by six months of systemic chemotherapy with agents F and O.

QUESTIONS

What patient-centered ethical concerns are raised by the discordance in treatment recommendations between the AI CDS system and human experts? How should disagreement between the system and the specialists be resolved?

How should Dr. King engage in shared decision-making with his patient? What information should the patient be given? Should the patient be informed of the relationship between the AI CDS system and the CEO of Premier Cancer Center?

CHAPTER FOUR

The Developer and AI Clinical Decision Support Systems

Developers are computer programmers who can write code and manage the technical and organizational elements of developing an algorithm. AI algorithms are sets of instructions for computers to perform specific tasks using predetermined data elements. Although AI algorithms learn from data, the process—even for unsupervised learning, in which algorithms generate their own outcome classifications—is guided by developers. Patients and providers value algorithm transparency, patient centeredness, and fairness, which can be achieved only when developers tune the data and the algorithm accordingly. Developers also have unique opportunities not only to ensure that algorithms meet minimum ethical standards, but also to generate algorithms that aid stakeholders in making ethical decisions by accurately representing the complex, nonlinear associations among bioethical principles and patient-specific factors (e.g., age, decision-making capacity), and then learning to identify bespoke, ethically sound solutions by performing thousands of scenario simulations (see figure 1 in chapter 1).

AI TRANSPARENCY

When AI algorithms entered the health-care domain, appreciation for their potential to offer predictive performance advantages was accompanied by reticence regarding their "black box" inner workings. Traditional statistical analyses and regression equations are familiar to most standard medical and premedical training programs, but these programs are only beginning to address the complex underlying mathematical concepts of AI. Thus, most health-care workers are ill equipped to interpret an AI model for their patients, who are often similarly unsure about how AI works. AI in health care suffers from a lack of explainability: transparency regarding how and why a model has generated a prediction or classification.

There are two schools of thought regarding AI explainability. One is that explainability is unnecessary. Acetaminophen provides excellent pain relief, and we incompletely understand how. Yet its risks and benefits are well established by decades of rigorous research and clinical experience; acetaminophen has earned trust. Conversely, trust in AI has been shaken by recognition that even highly accurate algorithms occasionally make egregious errors.[1] Until the safety and efficacy of AI algorithms are supported by large volumes of high-quality evidence, it seems prudent to endorse the other school of thought regarding AI explainabiltiy, lift the lid on AI black boxes, and understand their inner workings.

Explainable AI

Among all types of AI, deep learning is especially opaque. Difficulties in generating explainable deep learning are partially attributable to their architecture and design. Deep learning networks contain multiple layers of nodes (the computational units in a neural network) that are connected by mathematical functions. Each node has a weight that is influenced mathematically by the nodes in the previous layer, affects the output transmitted to the next layer, and changes over time as the algorithm learns associations between the input features and the outcome of interest. Heavier weights exert greater effects; lighter weights minimally affect the next layer. The final layer of the neural network compiles the final weights to produce a single output; for health-care applications, this output can be a patient's probability of having a heart attack or the classification of a skin lesion as benign or malignant. Although there are linear associations between nodes, the global process is deeply nonlinear. Therefore, it is difficult to represent mathematically how any single input feature affects the final prediction or classification.

One of the most promising and widely used methods for explaining AI is SHAP (SHaply Additive exPlanations).[2] SHAP approaches select parts of the dataset being used to train the algorithm, replaces the nonselected parts with values drawn from a similar dataset that will be used to test the algorithm, and then compares model predictions with and without each input feature. This process generates scores representing the effects of each feature on the final prediction (e.g., higher body temperature strongly predicts infection, the absence of diabetes weakly predicts the absence of infection). Importantly, SHAP values can be calculated across an entire dataset to show the global effect of a feature and can also be calculated for a single patient within a dataset, for personalized decision support. SHAP values may be useful not only in understanding how and why a model generated a prediction, but also in determining whether the learning process is consistent with medical knowledge. If a model learns to predict infection by linking socioeconomic status with infection, then developers should be skeptical and closely examine the training data and algorithm architecture for sources of bias. In some cases, the investigation may reveal that the association between socioeconomic status and infection exists, underscoring a greater problem: AI models usually lack causal inference mechanisms—means of understanding which features caused the outcome.

Associations, when considered in isolation, can deceive. After severe traumatic injury with ongoing bleeding, the early performance of procedures to stop the bleeding is necessary for survival. Therefore, one might reasonably expect to save lives by quickly transporting severely injured patients to a hospital that specializes in procedures to stop the bleeding. In one important study, the opposite was observed; shorter transport times were associated with increased mortality rates.[3] Deeper analysis of the study design and results revealed that many of the early transport patients were so severely injured that they probably would have been pronounced dead on scene if the emergency medical personnel hadn't responded so swiftly and rushed the patient to a hospital during the final few minutes of the patient's inevitably shortened life. Yet if an AI model trained on this dataset, it would "learn" to associate short transport times with increased risk for death. This issue is addressed by causal inference.

Causal AI

Causal AI answers the question, "Which features caused the outcome?" Most AI and standard statistical models derive associations and correlations between observations and outcomes and then predict outcomes on new, previously unseen observations. This approach is adequate, and sometimes optimal, when accurate predictions are the only desired product, but it informs decision support loosely and sometimes erroneously. An AI model predicting the timing of sunset may learn to *associate* falling atmospheric temperatures with imminent sunset, but an astute fifth grade student knows that cooling the atmosphere does not *cause* the sunset. In contrast, mechanistic models that use mathematical ground truth expressions (e.g., laws of gravity and planetary motion) to estimate the probability of an outcome can establish cause-and-effect relationships. In predicting how fast an object will free fall, we invoke the mathematical expression that gravity accelerates objects at 9.8 meters per second squared and (ignoring other forces like air resistance inducing terminal velocity constraints) apply a small set of causal predictor variables (initial velocity and height). This mechanistic approach allows causal inference by changing individual variables and assessing how those changes affect outcomes, as demonstrated in drug discovery and disease phenotyping applications.[4]

Uncertainty Awareness

When applying AI to an individual patient, building trust extends beyond explainability and causal inference, into the realm of uncertainty awareness. Can AI tell us the probability that a single point prediction or classification is accurate? Even models that usually perform well can make egregious errors, like a computer vision algorithm failing to see the elephant in the room (literally). Patients and clinicians may choose to ignore recommendations from a high-performing AI model, with understandable reticence that at any time the model could be making a rare but major error. Can an AI model that is correct 99 percent of the time alert us when it is making a one in one hundred mistake?

The challenge in quantifying the certainty of an AI prediction for a single patient is that common statistical measures of variance, like standard deviation and interquartile range, are undefined for point predictions: the outputs of AI algorithms for a single patient (e.g., predicted thirty-day mortality = 1.3%). To address this challenge, one may generate a series of point predictions to measure variance across the distribution (e.g., the set of predictions [1.3%, 1.2%, 1.4%, and 1.3%] has low variance and the set of predictions [1.3%, 3.5%, 0.1%, 1.9%] has higher variance). Yarin Gal and Zoubin Ghahraman propose an elegant solution for quantifying certainty in deep learning predictions.[5] When training a deep learning algorithm, it is often advantageous to drop different randomly selected sets of nodes with each pass of information through the model so that the algorithm avoids overfitting—finding perfect associations between the input features and the outcome of interest (near-perfect associations are problematic because the same near-perfect associations almost never exist in new, real-world data to which the trained algorithm will be applied); it is more effective for the algorithm to learn generalized, universally true associations that predominate in real-world data. Applying the same strategy (dropping randomly selected sets of nodes) *during testing* generates a series of deep learning models with slightly different architectures, thereby generating a series of point predictions. If those point predictions are tightly grouped (low variance and entropy), then model certainty for that prediction is high; if the point predictions are scattered widely (high variance and entropy), then model certainty in that prediction is low. By presenting AI-enabled decision support to users with uncertainty estimations, it may be possible to identify predictions and recommendations that are made with high confidence and low confidence. Therefore, like causal AI and explainable AI, uncertainty awareness can improve transparency. In turn, transparent AI has the potential to achieve the overarching goal of gaining the trust of patients and health-care providers.

DEVELOPER PERSPECTIVES ON PATIENT-CENTERED AI

Patient values are paramount in patient-centered decision-making and, more broadly, in health-care systems that respect individual autonomy. Yet patients and providers frequently misunderstand one another, and patient values are often overlooked by the prediction models and decision support systems that could remedy those misunderstandings.[6] Representing patient values in patient-centered AI is hindered by the inherent challenges in quantifying values. It may be possible to overcome those challenges. One potential solution is to incorporate the global preferences of target patient populations as model input features (while ensuring that preferences are measured equitably to accurately represent vulnerable patient populations). Another potential solution is multitask learning to generate a set of predictions or recommendations encompassing a range of potentially desirable outcomes to which patients may apply their own value systems in making well-informed choices.

Quantifying Patient Values

The global preferences of target patient populations can be incorporated in AI models as input features when such preferences are established by prior research studies. Consider a model recommending blood-thinning medications for a patient with atrial fibrillation—an abnormal heart rhythm in which clots can form within chambers of the heart and then travel to the brain, causing stroke. Such a model must balance risk for bleeding from blood-thinning medications against risk for stroke from blood clots. Those two competing risks are quantifiable from observations in research studies, but are bleeding and stroke equally detrimental? To answer this question, we must move beyond predictive analytics and ascertain patient values. Fortunately, researchers have published several high-quality studies about patient values regarding stroke and bleeding. Overall, patients consider one stroke to be equivalent to approximately five episodes of serious bleeding.[7] These results can be applied to known frequencies of bleeding among patients on blood thinners and then built into decision support systems. Unfortunately, high-level evidence regarding global preferences across target patient populations is unavailable for most day-to-day healthcare decisions; in these cases, it may be necessary to provide patients with a set of predictions or recommendations to which they may apply their own value systems.

Multitask Learning

Developers can incorporate individual patient-level values in AI-enabled decision support by using a multitask learning approach in which AI models generate sets of predictions for several different outcomes using one set of input features (e.g., simultaneously predict the probabilities of stroke, bleeding, and thirty-day mortality). This approach represents an alternative to training several individual models to predict each individual outcome. From a modeling perspective, multitask learning optimizes efficiency by avoiding the redundancy of applying patient data to several different models. Instead, patient data are applied to a single model, and the resulting data representation is passed through several outcome branches (one for each outcome being predicted) to produce prediction scores for each outcome, interpreted as the probability of that event occurring. In addition, and perhaps more importantly, multitask learning can improve predictive performance by allowing information gained from one prediction to sharpen other predictions. For example, a high probability of myocardial infarction is associated with a high probability of heart failure, so a multitask model predicting both outcomes simultaneously may be better at predicting heart failure than a model learning to predict heart failure in isolation.

For patient-centered AI, the greatest disadvantage of multitask learning is the implied requirement for patients to understand and interpret several model outputs according to their own values. The accompanying cognitive load and requirement for health literacy are substantial, though clinicians could ease these burdens by interpreting model outputs in clinical context with the wisdom of experience. There are also notable advantages to multitask learning. Providing sets of predicted probabilities for patient outcomes corresponding to specific care paths (e.g., surgery or no surgery) avoids making potentially erroneous assumptions about individual patient values and avoids the potentially coercive nature of providing treatment recommendations directly (e.g., "do this"). An aggregate patient population may think that one stroke equals about five episodes of serious bleeding, but a patient who wishes to avoid allogenic blood transfusions may place a higher priority on avoiding bleeding. By providing the predicted probabilities of both events rather than a recommendation to take or not take a blood thinner, the patient is empowered to apply their own value system and make a well-informed choice.

DEVELOPER PERSPECTIVES ON AI BIAS

AI can serve as a bulwark against biased decisions by anchoring the decision-making process in objectivity, but when applied improperly, AI can also potentiate implicit bias and social inequity in health care. Training an algorithm on imbalanced datasets that fail to represent target patient populations and using input features that have insidious associations with outcomes can each lead to biased AI predictions and recommendations. Identifying and mitigating the root causes of these sources of bias and inequity can tilt AI-enabled decision support toward justice and fairness.

Building Balanced Training Datasets

Algorithms trained on biased datasets can't help but produce biased outputs. Sometimes the source of bias is obvious. The initial phases of the Framingham Heart Study primarily enrolled White subjects. Associations between cardiovascular risk factors and events differ by race; an AI algorithm that learns to predict cardiovascular events from Framingham Heart Study data may generate biased predictions for a patient whose race is underrepresented in the dataset.[8] This source of bias is potentially amenable to ensuring that AI training datasets are built from broad, heterogenous patient populations that are representative of the patients for whom the AI application will be used, which is easy to say and difficult to do. Several elements contributing to imbalanced datasets exist upstream of developers. Research datasets in general, including those used for AI training, disproportionately represent socioeconomically advanced, cisgender, White men, who tend to seek care at well-resourced academic medical centers and consent to participation in medical research.[9] For groups that are underrepresented, AI models have insufficient data from which to draw sound conclusions and can produce unreliable recommendations.

Another threat to building balanced AI training datasets is reporting bias, by which we tend to collect data on patients who have received a treatment of interest. Returning to the example of high-risk surgery, reporting bias may decrease the representation of patients who were declined for surgery because they were at high risk for intraoperative and postoperative complications. When training datasets are built primarily using the lower-risk patients who underwent surgery, the resulting algorithm may have poor performance in predicting risk for the same high-risk patients who already suffer from poor access to care.

Developers can seek to mitigate these sources of bias by using real datasets to generate simulated datasets in which sociodemographic factors are representative of target patient populations (i.e., the simulations would decrease the proportion of overrepresented subjects while increasing the proportion of underrepresented subjects until the dataset matches general populations or the target patient population). At least one aspect of training dataset bias is refractory to this solution: patients without access to health care leave no data trail and therefore nothing to simulate. This problem is beyond the reach of the developer and falls to policymakers and society as a whole. Charles Binkley and David Kemp propose an "Access Pledge" for high-risk surgery whereby leaders at major health-care centers would promise to accommodate "anyone anywhere who wanted to come to their center for complex surgery."[10] In addition to its humanistic qualities, this approach concentrates high-risk surgery at centers that are most capable of caring for patients with complex surgical diseases.

Choosing the Right Model Input Features

Even when AI training datasets are balanced, algorithms may produce biased outputs if biased input features make important contributions to predictions. As discussed in the "Explainable AI" section, associations between model input features and the outcome of interest can be deceiving. This phenomenon has important implications for AI bias. If a decision support tool incorporates the observation that Black patients have increased risk for mortality after cardiac surgery, then model outputs could discourage Black patients from undergoing a surgery that would likely improve their health.[11] Input feature selection can also affect the accuracy of disease recognition and classification. Standard race adjustments for estimating kidney function underrepresent the prevalence of acute kidney injury among Black patients; in one investigation using AI models to identify acute kidney injury, race-agnostic models (those that ignore race) outperformed standard models that included race as an input feature.[12] Conversely, it is advantageous to include sociodemographic features in some AI models. When predicting hospital readmission after sepsis (infection with organ dysfunction), predictive performance is improved by including social determinants of health, and it is plausible that social factors (e.g., social support systems, food insecurity, the ability to fill prescriptions at a pharmacy) determine whether a patient remains well at home while recovering from a life-threatening infection.[13] Therefore, a blanket policy of excluding all potentially biased features from AI models would be suboptimal.

Emerging evidence suggests that developers may soon have an effective method for choosing unbiased features for AI models: mechanistic models. As further discussed in the "Causal AI" section, mechanistic models can identify features that *caused* the outcome of interest, rather than simply identifying features that were *associated with* the outcome. Black race may be associated with increased risk for mortality after cardiac surgery, but a mechanistic model could show that being Black does not cause death after cardiac surgery and could generate a list of other input features that do. Thus, mechanistic modeling, which requires substantial developer expertise as well as domain knowledge regarding mathematical ground truth expressions, has the potential to provide a data-driven method for deciding whether to include a potentially biased input feature in an AI model.

ENCODING BIOETHICS

By optimizing AI transparency and patient centeredness while minimizing bias, developers have unique opportunities to ensure that AI health-care applications meet ethical standards. Can AI move beyond the minimum requirements for meeting ethical standards in health care and help us make ethical decisions? Can AI encode bioethics?

Logical Expressions of Ethical Outcomes

To encode bioethics, AI must be confined to supervised learning by using outcomes that are defined by humans rather than patterns in the data, as in unsupervised learning. Defining ethical outcomes is challenging; defining ethical outcomes for AI is nearly impossible. Algorithms use formal, mathematical expressions exclusively. Patients' preferences for treatment options can be represented mathematically as rank-order lists, but such lists lack the qualitative nuances of real-world preferences. A preference for a high-risk palliative operation may arise from the desire to survive long enough to attend a granddaughter's wedding; the preference for surgical treatment is quantifiable, but the underlying value is not. Stripping away details also sacrifices some elements of truth, as George Box famously noted: "All models are wrong, but some are useful." Even some of the most successful models, like Newton's Laws, have limitations at certain scales. Defining ethical outcomes with logical expressions will inevitably sacrifice some degree of truth.

Logical Expressions of Ethical Principles

Defining ethical outcomes for AI is hindered not only by requirements to quantify values, but also by the prospect of encoding justice. Other major ethical principles in health care—autonomy (the patient's right to make their own decisions), beneficence (acting in the best interest of patients and society), and nonmaleficence (doing no harm)—are quantifiable in the sense that we can assign mathematical weights that represent their importance in a discrete decision-making scenario for an individual patient, but justice (equity in health-care delivery and outcomes) cannot be quantified fully for discrete decisions. Justice is partially quantifiable for certain scenarios under certain restrictions, like quantifying the risk of mortality and the projected costs of admitting a patient to an ICU, while recognizing that one less ICU bed will be available for all other patients, some of whom may have greater need of it. In this scenario, value of care (outcomes/cost) can be calculated for each patient, and justice regarding ICU bed allocation could, theoretically, be achieved by assigning the bed to whichever patient will gain the greatest value of care from ICU admission. Even then, one must assume that value of care has equal weight for each patient and is the sole contributor to justice in the decision-making process. What if one patient is likely to succumb to their illness even if they get the ICU bed (so value of care would be low for ICU admission), but their death would be less painful and more conducive to loved ones saying final goodbyes if they were in an ICU with continuous titrations of comfort care medications and a more spacious room to allow visitors? Therefore, contemporary solutions for encoding bioethics must operate primarily within the more limited realm of autonomy, beneficence, and nonmaleficence.

Ethical Frameworks for Encoding Bioethics

Given the challenges in defining justice with logical expressions, it may seem attractive to encode bioethics by abandoning the framework of autonomy, beneficence, nonmaleficence, and justice—the principles of biomedical ethics as originally described by Beauchamp and Childress—and pivot toward a different, classical framework.[14] Yet there are even greater challenges for encoding a consequentialist (utilitarian) ethical framework that values the best overall result, a deontological framework concerned with conforming to ethical norms, or a virtue framework concerned with personal character. The health-care domain knowledge required for a consequentialist framework would lead right back to the principles of biomedical ethics, the norms required for a deontological system are undefined for many health-care ethical dilemmas (unless one relies on the principles of biomedical ethics), and the virtue framework lacks the objectivity necessary for mathematical expression. Also, for encoding bioethics, it may be advantageous to use a comprehensive framework that encompasses all relevant domains of health-care ethics. Therefore, as endorsed by Meier and colleagues, we return to the biomedical ethical framework.[15]

Developer Implementation of Bioethical Code

The approach suggested by Meier and colleagues, which represents the state of the art for encoding bioethics, is based on fuzzy cognitive maps, which simulate complex systems (here, ethical decision-making) and learn from experience.[16] One of the advantages of this approach is that it allows for explicit encoding of defined elements, such as the principles of biomedical ethics themselves, as well as other salient, quantifiable elements, such as the patient's age and decision-making capacity. Other elements of the cognitive map must be derived from medical knowledge and research studies, such as the probabilities of gaining or losing length and quality of life. The developer then draws causal connections between nodes (e.g., the connection between having decision-making capacity and autonomy would be positive, and the connection between gaining length of life and high probability of death would be negative). Like standard deep neural networks, the nodes are connected by adjustable weights that are influenced by all incoming nodes and have the potential to affect subsequent nodes and the final model output. Unlike standard deep neural networks, fuzzy cognitive maps typically contain few nodes and can work with small datasets. The latter is fortuitous because large volumes of data are rarely available for the specific bioethical dilemmas (i.e., scientific investigations regarding topics like decision-making capacity and patient preferences for end-of-life care typically lack the situational granularity necessary for application to an individual patient). Deep neural networks also do not allow the developer to designate the meaning of individual nodes, as is done in fuzzy cognitive mapping. Herein lies another advantage for fuzzy cognitive maps for encoding bioethics: they have the inherent explainability that characterizes rule-based symbolic learning methods, while capitalizing on the deep learning ability to represent complex, nonlinear associations among elements. When Meier and colleagues leveraged fuzzy cognitive maps for encoding bioethics, the model agreed with biomedical ethicists 75 percent of the time. AI should not replace human expertise for a complex and sensitive application like resolving ethical dilemmas, but it does have the potential to encode bioethics and perhaps anchor ethical decision-making processes in objectivity.

SUMMARY

By writing code and managing the technical and organizational elements of developing an algorithm, developers—in close collaboration with patients and providers—can deliver transparent, patient-centered, fair AI algorithms in health care. Beyond ensuring that algorithms meet minimum ethical standards, emerging evidence suggests that developers can also generate algorithms that weigh ethical principles in decision-making paradigms. By these mechanisms, developers play critical roles in achieving the overarching goal of using AI to improve the health of individual patients and populations, while ensuring that such improvements are distributed equitably by serving as a bulwark against systematic and implicit bias. Like patients and providers, developers' role in ethical AI transcends individual applications and must be considered in the broader context of health-care systems and payers.

CASES

Case Study 4.1: Using Patient Demographics for AI Predictions

A seventy-year-old Black male visits his primary care provider (PCP) every few months to surveil and manage his chronic kidney disease and congestive heart failure. Despite reducing his dietary salt intake and taking his prescribed medications diligently, his renal function continues to decline. Given his history of complicated diverticulitis requiring multiple abdominal operations with extensive scar tissue formation, he is not a candidate for at-home peritoneal dialysis (using the abdominal cavity for dialysis). With worsening age-related macular degeneration superimposed on diabetic nephropathy, the looming eight-mile drive to the hemodialysis facility is troubling. He is also a loyal caretaker of his wife of fifty years, who recently suffered a debilitating stroke. Consistent with his wishes to avoid hemodialysis for as long as possible, his PCP endeavors to identify the latest possible date for safely initiating hemodialysis. She enlists a cutting-edge AI tool that predicts the timing of end-stage renal disease onset. Once end-stage renal disease appears imminent, the PCP will refer her patient to a surgeon for the creation of an upper extremity arteriovenous fistula—a connection between an artery and a vein in the arm—so that the fistula has time (six weeks or more) to mature and be used for hemodialysis, avoiding placement of a plastic dialysis catheter and its attendant risk for bloodstream infection.

One morning the PCP receives a notification that her patient has been admitted to the ICU with a heart failure exacerbation secondary to renal failure and volume overload. Usually she'd spend her lunch break reading peer-reviewed medical journals about recent advances in chronic disease surveillance and management; today, she drives to the hospital to visit her patient. He expresses gratitude for her visit in short, labored phrases, breathing heavily against the fluid accumulating in his lungs. A plastic dialysis catheter is placed for the emergent initiation of hemodialysis. He begins to recover but then quickly declines after developing a systemic infection from his dialysis catheter. Despite prompt recognition and appropriate treatment, his organs fail one after another, and he passes peacefully.

Two weeks later, during a solemn lunch break review of current literature, his PCP happens upon a scathing review of AI CDS, focusing on the inappropriate use of race as a prognostic variable. The article illustrates that when race is used to predict progression of chronic kidney disease, algorithms—like the one she adopted—underestimate disease severity for Black patients, resulting in errantly late recommendations regarding the timing of initiating dialysis. Conversely, race-agnostic algorithms consistently make accurate, equitable recommendations. Her appetite gives way to nausea.

QUESTIONS

Who is responsible for the prognostic errors that were built into the algorithm the PCP used?

Can the patients and physicians who use AI CDS be expected to anticipate methodological flaws in algorithm design?

After the prognostic error is disclosed to the patient's widow, how can her trust in the health-care system be restored?

Case Study 4.2: Ignoring Patient Demographics for AI Predictions

Dr. Vincent graduated in the top of her medical school class and was encouraged by many to pursue a prestigious, high-paying career as a procedural subspecialist in a big city. While she appreciated this well intended advice, she chose instead to serve patients in a community setting as a PCP in rural Wyoming, diagnosing and treating a wide range of conditions while promoting healthy lifestyle choices.

One of the challenges in providing excellent primary care without dependence on specialists is mastering the enormous scope of practice. Staying up to date on contemporary practices for managing common conditions like hypertension and diabetes is relatively easy, but staying up to date on thousands of rare conditions—any of which may affect her patients—is arduous, if not impossible. Therefore, Dr. Vincent augments her clinical workflows with an AI-enabled system that uses information captured in the EMR to generate a list of diagnoses that are highly likely to fit the patient and a list of diagnoses that are moderately likely to fit the patient but could be devastating if missed (e.g., shortness of breath is probably due to an asthma exacerbation, but could be due to pulmonary embolism—a blood clot in the lungs that is occasionally fatal). Usually, Dr. Vincent has already considered the diagnosis lists generated by the AI system; occasionally, the AI system raises the specter of a rare disease that is confirmed by subsequent diagnostic testing. Dr. Vincent's patients know that she uses the system and checks its outputs carefully against her own independent assessment. All parties learn to trust and appreciate the AI system's use within this context.

In response to federal guidance extending antidiscrimination laws to clinical algorithms, the AI developer strips age and sex from the EMR data that the AI system uses to generate diagnosis lists. In many cases, losing these input variables has minimal effect on system performance. For conditions that have strong associations with age or sex, system performance plummets. Dr. Vincent's patients still garner the full benefits of her clinical expertise, but in some cases they are no longer garnering the full benefits of the AI system.

First, Dr. Vincent misdiagnoses joint pain as osteoporosis, when the true etiology is an atypical presentation of rheumatoid arthritis, an autoimmune disorder that affects joints. Within months she notes that her patient is responding poorly to her prescribed treatment and successfully changes course to target the lurking autoimmune disease, but she is frustrated that her patient has endured months of potentially avoidable pain and suffering. Then Dr. Vincent misdiagnoses chest pressure as anxiety, when the true etiology is an atypical presentation of a heart attack. This error is not identified at a clinic visit months later; it is identified when her patient becomes unresponsive at dinner, prompting a 911 call. One of the few county ambulances happens to be nearby, and the swift actions of the emergency medical service (EMS) crew saves her patient, who then undergoes emergency stenting of the arteries in his heart and remains ill in an intensive care unit for almost two weeks. Dr. Vincent wants the old AI system, and its potentially discriminatory age and sex input variables, back.

QUESTIONS

Should patients and physicians tolerate the potential for AI systems to discriminate, if doing so improves their performance and improves patient care?

If it's wrong for AI systems to consider variables like age and sex when making diagnoses, should physicians be trained to ignore these variables as well?

Case Study 4.3: Developer Incentivization Targets (What to Optimize—Payer/Health System Goals or Patient goals, Which May Conflict)

Ms. Januz is a talented, hard-working developer and entrepreneur who is the CEO of a small but vibrant health-care AI start-up company. Her multidisciplinary team comprises experts from clinical, computer engineering, and implementation science domains who are passionate about developing AI CDS that improves health care. Her company has developed a competitive advantage with their plug-and-play AI system that predicts how patients will fare if they undergo a major surgery and uses the goals and values of individual patients to guide system recommendations. The AI system is increasingly used by both patients and physicians who are involved in shared decision-making processes regarding major surgery. In addition to the positive sentiments shared by patients and physicians who have used the system, three independent investigator groups have performed scientific experiments demonstrating that use of the AI system is associated with decreased decisional regret and greater patient satisfaction. Ms. Januz receives calls and emails every day from hospital administrators who are interested in purchasing and implementing her software.

As her company grows exponentially, one of the largest health-care systems in the United States offers to fly Ms. Januz and her team to its national headquarters. After a warm welcome and fancy hors d'oeuvres, the main agenda item is brought forth: the company wants to purchase and implement the AI system in hundreds of hospitals and clinics, but only if it can be certain that doing so will not adversely affect the incidence of mortality within ninety days of surgery, which is a primary performance metric. Since the system uses patient goals and values to generate recommendations, it isn't designed to ensure that patients survive for ninety days. For a patient with a life expectancy of two to three months due to cancer in the pancreas that is obstructing the flow of food out of the stomach and bile out of the liver, palliative surgery that bypasses the obstruction can provide major relief of symptoms but is not intended to prolong life. In contrast, offering the palliative surgery incurs high risk of mortality within ninety days of surgery. Yet if patients' values are considered, many will opt to undergo palliative surgery and minimize their suffering from the nausea and itchiness that accompany obstruction of the stomach and liver. Therefore, Ms. Januz is asked to minimize the relative importance of patient values in her AI system and shift its focus toward minimizing ninety-day postoperative mortality. She has two weeks to decide whether to comply, accept a contract that would triple her company's revenue overnight, and know that the modified system would still provide highly accurate predictions of postoperative outcomes that could augment shared decisions regarding major surgery. Or she can maintain her company's commitment to incorporating patient values in the AI system and remain on a course that has been successful thus far, while reaching fewer patients with her software.

QUESTIONS

Should Ms. Januz modify the AI system as requested? Why or why not?

Can patient autonomy be respected without encoding patient values in AI CDS?

CHAPTER FIVE

The Health System Executive and AI Clinical Decision Support Systems

INTRODUCTION

Health system executives—payers (insurance companies) and administrators—negotiate financial incentives and exercise oversight of health-care quality and safety, thereby wielding substantial authority over health-care delivery. Ideally, incentives and oversight-related regulatory actions are grounded in reliable evidence from well-designed scientific experiments that undergo rigorous peer review, but high-level evidence for AI-enabled decision support is sparse. While researcher teams of physician scientists and developers seek to fill clinical AI knowledge gaps, health system administrators and payers must negotiate a balance between the perils of early, hasty implementation—which could cause preventable harm and further erode trust in AI—versus timid, late implementation that fails to capitalize on AI's potential to improve patient safety, make operational structures more efficient, contain rising costs, and even reduce physician burnout. The advantageous, middle ground is characterized by thoughtful incentivization, vigilant oversight, and epistemological modesty—investigation of that which separates justified belief from opinion. By overseeing AI CDS system implementation and reimbursement, health system executives define the long-term goals of developers and exert high-level influence on the daily experiences of patients and physicians.

EXECUTIVE-DRIVEN FINANCIAL INCENTIVES

While researcher teams—typically comprising physician scientists and developers—are responsible for generating evidence regarding AI CDS systems, health-care administrators and payers must interpret evidence and balance competing risks of early versus late implementation. While physicians often enjoy a singular focus on providing the best possible care for individual patients, executives must determine which health-care interventions are allowed and reimbursed for millions of patients while facing intense scrutiny for decisions that, if errant, could quickly harm many.

Balancing Risks of Early versus Late Clinical Implementation

Early, hasty implementation of AI CDS could create preventable harm and further erode trust at a time when building trust in AI is critical; late, timid implementation could fail to achieve the potential beneficence of AI for several important goals: improving patient safety through prevention and early identification of errors and patient deterioration events, making operational structures more efficient by streamlining supply chains and more precisely allocating resources, and reducing physician burnout by automating rote tasks. Executives must balance these competing risks and find the optimal, middle ground. Health care has tended toward late implementation; the average time lag between the publication of original research and clinical implementation or inclusion in guidelines ranges from nine to twenty-eight years and averages seventeen years.[1]

Health-care industry leaders and informed citizens occasionally criticize the United States Food and Drug Administration (FDA) for being overly cautious and thus delaying the arrival of new interventions at the bedside. The FDA, formerly responsible only for supervising the safety of food, medication, and medical devices, now maintains a software as medical device category to which AI CDS belongs. As noted by Alex Tabarrok, delayed FDA approvals of health-care products cost lives, but lives are taken by a disease rather than a medical treatment, and so "the bodies are buried in an invisible graveyard." In contrast, lives taken by an unsafe product are sins of commission—the product itself kills people, whose stories are shared by news agencies and social media outlets, and whose loved ones may take legal action and receive compensation. Consider the short- and long-term theoretical consequences of an FDA-approved COVID-19 vaccine causing more harm than good. While COVID-19 ravaged millions worldwide, pharmaceutical company leaders pledged to apply for FDA approval *only after* rigorous trial data were already available; this role reversal, in which pharmaceutical companies self-imposed a delay in the development-to-market pathway, illustrates the risk inherent in executive-level clinical implementation decisions.[2]

There are computational solutions for identifying the optimal balance of aggression and conservation in FDA approval based on randomized controlled trials, the gold standard study design for health-care interventions. Randomized trials balance the probability of a type I error—an intervention falsely appears to help—against the probability of a type II error—an intervention falsely appears to make no difference. The FDA could achieve a type I error rate of 0.0 percent by approving nothing, or a type II error rate of 0.0 percent by approving everything. The traditional approach to finding the optimal, middle ground is to approve interventions in which a randomized trial generates a *p* value—here, the probability that the observed difference in outcomes would occur if the intervention truly had no effect—of 5.0 percent or less (i.e., low probability that the observed difference in outcomes was due to random chance). Once that threshold is set, the type II error rate is calculated according to the sample size (usually the number of patients in the study) and the anticipated treatment effect according to previous studies, and typically ranges from 10 to 20 percent. But are these arbitrary thresholds always optimal? According to FDA experts, the answer is no: "Alternative values to the conventional levels of type I and type II error may be acceptable or even preferable in some cases."[3] So how does one determine these elusive, alternative values?

Leah Isakov, Andrew Lo, and Vahid Montazerhodjat describe a Bayesian decision analysis approach—a probability-based framework for formally evaluating a decision—that calculates optimal thresholds by assigning *context-specific costs* to type I and type II errors.[4] This approach allows tailoring of thresholds to the intervention and target disease. If established interventions are lacking and the disease kills previously healthy people quickly, the cost of falsely disproving the efficacy of a new, beneficial treatment is high; if effective interventions already exist for an indolent disease that mildly worsens quality of life, the cost of errantly approving a new, ineffective intervention with dangerous side effects is high. By incorporating contextual elements in a formal mathematical solution, Isakov and colleagues demonstrate that standard thresholds for FDA approval are far too conservative when considering deadly diseases with no viable alternative intervention and are too aggressive when considering indolent diseases. Thus, the Bayesian decision analysis could function as decision support in FDA approval processes, toward balancing risks of early versus late clinical implementation.

Evidence-Based Incentivization

FDA approval is a necessary, early step toward clinical implementation of new health-care interventions; payers exert subsequent control, with substantial effects. Angioplasty and stenting—using balloons and expandable tubes to hold arteries open—is a standard treatment for stenotic (narrowed) arteries in the heart, which can cause heart attacks. In 2005, the FDA approved humanitarian use of angioplasty and stenting for intracranial stenoses (narrowed brain arteries) that cause strokes.[5] The following year, the Centers for Medicare & Medicaid (CMS) determined that intracranial angioplasty and stenting would be reimbursed only within a randomized controlled trial. In 2008, despite building industry pressure to push financial incentives forward, which would drive widespread clinical implementation, CMS insisted that only high-level evidence would earn incentivization.[6] Three years later, the SAMMPRIS trial showed that angioplasty and stenting of intracranial stenosis caused a threefold increased risk of stroke or death.[7] AI CDS, like procedural interventions or medications, has the potential to demonstrate efficacy in observational studies but subsequently prove ineffective or harmful in randomized controlled trials that minimize or eliminate the experimental sources of bias that plague observational studies. Thus, AI CDS should meet the same high evidence-based standards as procedural interventions and medications.

AI CDS, compared with other health-care diagnostic and therapeutic interventions, has a short track record of support from randomized controlled trials. In 2017, researchers reported that an AI-enabled tool for predicting sepsis (life-threatening infection) among critically ill patients led to shorter length of stay in the hospital and decreased in-hospital mortality.[8] In the subsequent and aptly named HYPE trial, an AI CDS system predicted intraoperative hypotension and prompted anesthesiologists to act earlier, more often, and differently, resulting in fewer hypotensive episodes during surgery and less overall time-weighted hypotension.[9] Although other, similar trials have demonstrated success for AI CDS, and there are surely more successes to come, other strategies are needed to carry scientific investigation forward. Randomized controlled trials remain the gold standard for establishing causal relationships, supporting reimbursement, and driving widespread clinical implementation, but they are costly, time-consuming, and often performed at ivory tower academic institutions, enrolling patient populations that do not accurately represent the geographic and socioeconomic diversity of the patients for whom the intervention will ultimately be used; alternative study designs are needed.[10]

For generating high-level evidence to support AI CDS, pragmatic and digital trials offer attractive alternatives to traditional randomized controlled trials. Pragmatic trials leverage existing clinical environments and workflows as backdrops and mechanisms for delivering the intervention.[11] The internal consistency of delivering the intervention and the circumstances under which it is delivered are less controlled and potentially compromised, but they bend toward real-world delivery and circumstances, thus avoiding the pitfall in which an intervention is effective under the tightly controlled conditions of a randomized trial but fails when implemented broadly alongside standard clinical care. Pragmatic trials also harbor the potential to help executives understand whether interventions improve value of care (patient outcomes / health-care cost) and trigger de-implementation of low-value care, like magnetic resonance imaging for low back pain. AI performs well in generating predictions used in observed-to-expected patient outcome ratios for health-care quality and value benchmarking; delivering value of care predictions to health-care executives could inform implementation and de-implementation decisions. Digital trials leverage existing data from EMRs, mobile devices, wearable sensors, and in-home sensors to generate robust, holistic representations of patient health without requiring the cost, infrastructure, travel, or time to make separate measurements of patient health indicators (i.e., vital signs, laboratory values) solely for research purposes.[12] When necessary or advantageous, digital trials can be completed entirely virtually, as demonstrated during the COVID-19 pandemic.[13] Like pragmatic trials, digital trials sacrifice internal consistency in favor of real-world circumstances.

Combining pragmatic and digital trial concepts has the potential to efficiently and effectively generate high-level evidence to support AI CDS, allowing health-care administrators and payers to make informed decisions regarding incentivization. Only with a strong evidence base can executives achieve epistemological modesty—honest, sincere investigation of that which separates justified belief from opinion. This often requires a shift away from AI hype in favor of a more tempered mindset regarding the potential for AI to improve clinical decision-making and patient outcomes, toward wise incentivization decisions and appropriate oversight.

EXECUTIVE OVERSIGHT

Accountability for the effects of AI-enabled decision support lies not only with the physicians who are responsible for the veracity and efficacy of clinical decisions, but also with the health system administrators who choose whether to integrate AI-enabled decision support with established digital workflows at large scale. Beyond the individual patient-physician interactions with AI CDS that are uniquely personal, health system administrators must maintain a higher-level view of systemic effects, prevent potentially harmful scenarios, surveil for performance degradation, and protect patient data from unintentional disclosure.

Preventing Harm

After the space shuttle *Challenger* tragedy, the National Aeronautics and Space Administration (NASA) adopted technology readiness level assessments that indicate whether a new technology is ready to fly into outer space with human lives at stake. Articles on AI CDS systems are more likely to be published in peer-reviewed journals when they produce impressive, highly accurate predictions; research teams of physician scientists and developers, internally and externally incentivized to publish important scientific discoveries, may hesitate to illustrate instances in which the model performs poorly. Therefore, executives—unbound by similar pressures—should insist that AI CDS systems be subjected to technology readiness level assessments that indicate whether they are ready to integrate with clinical care.[14]

Preimplementation technology readiness testing could include simulation of rare and erroneous model inputs with assessment of how model performance is affected and why failures occurred.[15] Subsequent investigation could mimic drug discovery phases 1 and 2 clinical trials by performing prospective studies under close surveillance and high scrutiny, then replacing traditional phase 3 and 4 clinical trials with pragmatic, digital trials, as described in the section "Evidence-Based Incentivization." Only when an AI CDS system demonstrates safety and efficacy throughout this continuum of scientific investigation should it be disseminated and implemented broadly. Even then, AI CDS requires continual surveillance.

Auditing for Harm

Even after rigorous testing of AI CDS systems with technology readiness assessments and pragmatic, digital trials, postimplementation surveillance is necessary. When models learn a close fit between input variables and outcomes in training data, shifts in those associations over time will compromise performance for locked models that are no longer learning or relearning. This phenomenon, termed *dataset shift*, is a major threat to postimplementation AI CDS.[16] In the wake of the COVID-19 pandemic, shifting associations between fever and bacterial sepsis were adversely affecting sepsis alerts generated by a prominent AI-enabled sepsis prediction model.[17] This cautionary event illustrated the importance of surveillance; a clinical AI governing committee noted the issue and decommissioned the CDS system expeditiously. In contrast to the COVID-19 pandemic, many causes of dataset shift are subtle; patient demographics, hospital policies, and measurement instruments change over time, and each change may degrade model performance insidiously. Surveillance of model performance by governance committees is essential for early detection of dataset shift and should be overseen by executives. In parallel, executives may consider incentivizing developers to implement online learning models that, unlike locked models, continue to learn associations between inputs and outcomes over time. While online learning may mitigate risk for performance degradation due to dataset shift, the intensity of surveillance required for online learning eclipses that of locked models because at every time point, the "live" version of the online model is new and untested. In keeping with the theme of doing no harm, executives should also oversee policies and procedures governing data security and privacy.

Data Security and Privacy

Health-care administrators must ensure that AI CDS systems comply with national and international oversight regarding data security and privacy (e.g., HIPAA in the United States and the General Data Protection Regulation [GDPR] in the European Union). The HIPAA Privacy Rule safeguards protected health information (PHI), which it defines as information that can be explicitly linked to a particular individual. Researchers can maintain compliance by redacting all PHI by the safe harbor method, or by obtaining local institutional board approval to receive a limited dataset that contains some PHI (usually dates and locations are preserved, while more sensitive identifiers, like names and dates of birth, are redacted), or by expert determination, in which one applies accepted statistical and scientific de-identification methodologies to the dataset, such that the risk of identifying an individual is very small. Compared with HIPAA, the GDPR imposes tighter restrictions on the use and disclosure of personal data, which it defines broadly as anything that could be used to identify an individual. Under GDPR, individuals may access and correct their own data, know where their data originated, opt out of automated decision-making, exercise the "right to be forgotten" (i.e., mandate that their existing data are erased), and withhold permission to use their data. Even after permission is granted, which must be done by unambiguous consent, there are restrictions on subsequent downstream transfer of data to other institutions or entities.

Whether working under the dominion of HIPAA or GDPR, health-care administrators must ensure data security and protect patient privacy without compromising AI CDS performance by unnecessarily limiting data availability. Perfect data security could be achieved by banning all use and disclosure of health-care data outside of individual patient-physician relationships, but the ensuing lack of data availability would severely hinder scientific progress in general, and the development and implementation of AI CDS specifically. Federated learning can achieve balance between data security and availability.

In federated learning, local AI models train separately on local data. As they train, local models at partner sites send updates (model weights) for consolidation into a global model, which is then shared with partner sites. The process is repeated until model performance ceases to improve. In some cases, global models outperform local models on new, previously unseen local datasets, suggesting that acquiring more, diverse data for AI training datasets is a positive-sum game.[18] This performance advantage is likely attributable to learning more accurate representations from rare instances that are underrepresented in small datasets, a phenomenon that also has the potential to improve CDS equity across common and rare diseases and sociodemographic profiles.

Federated learning platforms maintain data security and privacy by sharing model weights rather than patient data. Communications between central and local models use authentication and encryption software and transmission protocols that allow messages to be sent securely over the internet. Each partner site can also implement its own local authentication, authorization, and privacy policies without communicating its policies with the server or other sites. Despite these safeguards, adversaries (hackers) can infer whether patients belong to the training dataset and can reconstruct model inputs from model outputs.[19] Adversarial attacks can be countered by intentionally embedding noise into input data, the model itself, or model outputs—although this approach can compromise model performance—and by clustering-based anonymization, in which information representing an individual in a batch of data cannot be distinguished from information representing other individuals in the same batch.[20] Collectively, the absence of sharing patient data and the presence of additional protections against adversarial attacks offer the ability to perform collaborative AI modeling across partner sites with minimal risk to data security and privacy. While the federated learning architecture and implementation requires the expertise of developers, health-care administrators should consider the advantages of approving and incentivizing federated learning methods in the development and implementation of AI CDS, thus simultaneously promoting data security, patient privacy, and data availability.

SUMMARY

Health system executives typically have skill sets that are tailored to health-care finance and business administration rather than computer programming or engineering. Yet payers and administrators are responsible for system-level decisions regarding implementing and reimbursing AI-enabled decision support, with substantial downstream effects on patients, physicians, developers, and scientific progress. It is imperative that executives make decisions and establish policies that support the safe, effective development and clinical implementation of AI CDS while communicating these decisions and policies to other stakeholders in a manner that promotes a shared vision and shared goals and objectives. By balancing the risks of early versus late clinical implementation; using evidence to guide incentivization; and exercising administrative authority to minimize harm and ensure data security, privacy, and availability, executives play essential roles in the safe, effective, equitable delivery of AI-enabled decision support.

CASES

Case Study 5.1: Predicting Futility in Massive Transfusion

A thirty-seven-year-old mother of four is the restrained front-seat passenger in a high-speed motor vehicle crash. When EMS personnel arrive, she is in shock from massive blood loss. Initial physical examination demonstrates a mangled right upper extremity with active bleeding from the upper arm at the site of a deep laceration, to which a tourniquet is applied, as well as an unstable pelvic fracture. The EMS team transfuses two units of red blood cells during transport to the hospital.

On arrival at the emergency department, the woman is unresponsive. A breathing tube is placed and attached to a mechanical ventilator by the emergency medicine team. A pelvic binder is applied to slow the rate of bleeding from pelvic fractures, and a rapid infusion device is used to transfuse approximately one liter of blood per minute, meaning that her entire circulating blood volume will be replaced with transfused blood every five to seven minutes. Abdominal ultrasound demonstrates a large volume of internal bleeding.

The woman is transported emergently to the operating room, where exploratory surgery identifies a large shearing injury of the inferior vena cava, the largest vein in the body, as well as major injuries to the liver and spleen. Pressure is applied to the vena cava and liver injuries while the spleen is removed. Exploration of the right upper extremity injury reveals a transection of the brachial artery—the main vessel carrying blood to the forearm and hand. A temporary shunt is placed to restore blood flow. The pressure applied to the liver has slowed the bleeding at that site, but there is persistent, massive hemorrhage from the vena cava, and the patient is increasingly unstable despite ongoing massive blood transfusion. She has now received forty units of red blood cells and thirty-four units each of plasma and platelets—the components of blood that aid in clotting—as well as multiple doses of medications that promote blood clot formation. The anesthesia team has also administered medications to normalize the acid-base status of the circulating blood and restore blood calcium levels, which are necessary for clotting. To stop the bleeding quickly, the vena cava must be ligated—closed with suture, rather than repaired. With this maneuver the bleeding slows but persists from the raw, cut surfaces of the surgical dissection and crushed pelvic bones. The blood bank calls to inform the anesthesia and surgical teams that an AI-enabled decision-support system has indicated that the probability of patient survival is now 0.00 percent. Therefore, the blood bank will no longer be sending additional red blood cells, plasma, and platelets to the operating room. Instead, those blood products, now in short supply, will be preserved. A spirited discussion ensues in which the attending anesthesiologist and surgeon agree that with ongoing resuscitation, the patient has a very small but real chance of survival and functional recovery. The blood bank, bound by evidence-based, executive-approved institutional policy, refuses to send more blood products to the operating room. As the patient's blood pressure falls, the trauma alert pager sounds off with a sobering report: EMS is en route with another severely injured patient who is becoming increasingly unstable and requiring red blood cell transfusion.

QUESTIONS

Are there conditions under which an AI CDS system can be entrusted with futility decisions?

Is it ethically sound for hospital system administrators to use AI CDS to ration care at the population level? Should payers incentivize the use of AI CDS systems that promote high-value care by restricting use of resources that are unlikely to improve outcomes, even in emergency situations?

How might executives set thresholds for humans to override AI recommendations? Should the thresholds be titrated to the probability of disagreement, the severity of consequences for a suboptimal decision, or both?

Case Study 5.2: A Tale of Two Payers—Level 1 Evidence versus Focused Empiricism in AI Research

Payer 1 has a clear and time-honored reimbursement policy for new health-care interventions: show us the level 1 evidence for it, and we will pay you. In contrast, payer 2 makes case-by-case decisions about reimbursing new interventions according to focused empiricism: using practical observations and the best available evidence when there is a lack of federal funding or it would be impractical to general level 1 evidence (which encompasses most new health-care interventions). Payer 1 rationalizes its policy by recognizing that level 1 evidence from randomized controlled trials remains the gold standard for investigations in healthcare. Payer 2 rationalizes its policy by recognizing that in many cases, physicians provide care that is believed to be optimal based on observational (nonrandomized) data because many effective treatments have not been tested in a randomized trial, but the weight of available evidence establishes their effectiveness—and it would be wrong to withhold these treatments just because it is too expensive or impractical to perform a randomized trial.

As expected, clinical management decisions follow reimbursement patterns. Therefore, patients insured by payer 1 are less likely to be subjected to a new health-care intervention than are patients insured by payer 2. Although some patients have financial opportunities to obtain insurance coverage from the payer of their choosing, most patients do not, and they are beholden to the reimbursement policies determined by the payer they can afford. This financial element exacerbates disparities in health care.

QUESTIONS

How do the reimbursement policies of payer 1 and payer 2 satisfy the bioethical principles of beneficence and nonmaleficence? Which payer achieves a more favorable balance of beneficence and nonmaleficence?

How does the association between patient socioeconomic status and payer affect the bioethical principle of justice?

Case Study 5.3: A Tale of Two Hospital—Early versus Late Implementation of AI CDS

Hospital 1 is a late adopter of new health-care interventions. After most other hospitals have implemented an intervention and found that it works well, hospital 1 slowly, deliberately implements the intervention by applying lessons learned by others along the innovation learning curve. In contrast, hospital 2 is an early adopter of new health-care interventions. As soon as focused empiricism suggests that an intervention provides a performance advantage or improves patient outcomes, hospital 1 invests substantial resources in using validated implementation science frameworks to guide the early implementation process. Hospital 1 develops a reputation as a science and technology laggard that delivers safe and effective care; hospital 2 develops a reputation as an innovation lightning rod that provides cutting-edge care that often evolves the standard of care and occasionally imparts harm from promising interventions with latent adverse effects.

Overall, the two hospitals have similar patient outcomes. The few instances in which the early adoption of a new intervention at hospital 2 fails are offset by the many instances in which the new intervention improves patient outcomes. Each time a new intervention succeeds at hospital 2, there are hundreds of patients treated at hospital 1 with the inferior historical standard of care, but these failures are less notable because the public knows and accepts that the standard of care is imperfect. When a new intervention fails at hospital 2 and patients were harmed or killed by a new treatment, this makes the front page of the news and a class action lawsuit ensues.

As the headline failures and lawsuits pile up, hospital 2 loses its appetite for innovation and sinks into a late adopter role. Other early adopter hospitals do the same.

QUESTIONS

As early adopter hospitals sink into late adopter roles, how will this affect late adopter hospitals and innovation in general? What are the downstream effects on patients?

Physicians take an oath to "first, do no harm," consistent with a late adopter approach. Is this best for health care, or should some physicians innovate, knowing that some harm will be caused to a few, well-informed patients so that outcomes can be improved for all other patients?

CHAPTER SIX

Incorporating Ethics into the AI Clinical Decision Support System Life Cycle

Considering AI CDS from the distinct perspective of each of the stakeholder groups is essential. As these systems are conceptualized, developed, deployed, utilized, and audited, ultimately the ethical concerns of each of the stakeholders will intersect, and in many cases overlap. There are both general considerations that are applicable to all phases of the AI CDS life cycle and some that are distinct and applicable to specific phases. This chapter begins with general considerations, then the ethically salient phases of the life cycle are defined, and finally the concerns raised during each phase are discussed.

GENERAL CONSIDERATIONS

A central theme of this book is that patients are at the center of health care and that it is through the physician-patient relationship that patients have additional relationships with healthcare systems, payers, and AI CDS developers. The importance of patients and the patient-physician relationship governs the ethical considerations of implementing AI CDS from concept to use. There are multiple frameworks, pipelines, and systems that suggest blueprints for the ethical integration of AI CDS.[1] This book's proposal for assuring that ethical concerns are identified and addressed at each stage in the AI CDS life cycle differs in important ways. First, as was highlighted earlier, this proposal is explicitly patient centric. In addition, it considers the other roles that the patient may occupy in relation to AI CDS, namely the patient as data donor, learning subject, and research subject. Second, it is collaborative and multidisciplinary. Third, it recognizes that there are very few current concrete regulations and requirements to ethically develop and implement AI systems to support clinical decisions, similar to other innovative technologies.[2] Thus, this proposal tries to provide both general guidance and concrete actions that are necessary for encoding ethics into AI CDS, favoring the latter.[3] Fourth, in keeping with a patient-centric perspective, the importance of defining thresholds for benefit and burden is emphasized. And last but of no less importance is the ever-present need to consider equity and to recognize the potential of these systems to perpetuate bias.

1. **Each health system and payer who plans to implement AI CDS should have an oversight group.** Throughout the life cycle of every AI CDS system, from idea and concept to use and auditing, there should be a multidisciplinary group of stakeholders that guide and oversee the process. There should be a core group of stakeholders who are involved in every project, with additional expertise added on an ad hoc basis depending on the relevant clinical decision. This is true whether the

system is developed internally or purchased from a commercial vendor.

At a minimum, the group should have representation from the following roles:

A. Patients, families, and community members
B. Clinical:
   a. Clinicians with applicable expertise in AI and/or medical informatics
   b. Clinical leadership, especially trusted clinical leaders
   c. Clinicians with clinical research experience
   d. Clinicians with experience applicable to the specific system being developed (may be ad hoc)
   e. Clinicians drawn from a variety of clinical disciplines and not entirely limited to physicians
C. Technology:
   a. Data science
   b. Human factors engineers
   c. Programmers
   d. Internal expertise, even if the systems are being purchased from vendors
D. Legal:
   a. Health and research law
   b. Intellectual property law
E. Quality and patient safety
F. Risk
G. Compliance
H. Bioethics
I. Diversity, equity, inclusion
J. Human resources

K. Patient experience

L. Research oversight/institutional review board (IRB)

2. **Patients, families, and community representatives should be part of the oversight group.** Having patients, families, and community representation as part of the group will inevitably bring insights and perspectives that otherwise would be lost. Of utmost importance, all involved in AI CDS must remember that it is for patients that these systems exist, and patients, families, and community members must not be viewed as tokens, or even more egregious, fortunate to be given an opportunity to participate in the process. They should be viewed and treated as valuable and essential voices in the AI CDS life cycle.[4]
3. **The oversight group should be genuinely diverse.** In addition to diversity in terms of disciplines and professional roles, a group such as this would benefit from diversity in terms of personal experience and rank within the health-care hierarchy. Using AI to support clinical decisions is largely uncharted territory. Diverse clinical perspectives and personal experiences will be very valuable to highlight areas of concern and address them before moving forward with the project.
4. **The oversight group should have a clearly defined role.** This includes the scope of the group's authority, the group's reporting structure within the larger organization, and the manner in which the group will communicate its recommendations to the organization. One model is that the oversight group acts as an authoritative body that has the power to make important, relevant, and

binding decisions if a majority of the members agree.[5] At the other extreme, the group would be an example of "ethics whitewashing," wherein oversight is symbolic and largely for the sake of appearances, with power concentrated in executive leadership.[6]

A reasonable compromise might be that the group has a wide scope of oversight along the AI CDS life cycle. The group is actively involved in all AI projects from inception and retains oversight and algorithmic stewardship.[7] Some members may play greater and lesser roles during the system's life cycle, but there would be group ownership of these systems until they were decommissioned. The group would report at regular intervals to executive leadership, including technological, clinical, and operational. Recommendations would be weighted and substantiated, based on research and expertise.

Differing perspectives would not require reconciliation but rather would be presented with their accompanying reasoning and perspective, allowing for informed decision-making at a higher organizational level. A well-assembled oversight group would be expected, even encouraged, to have differing opinions in a nascent field like AI CDS. From a meta perspective, much like clinical decisions themselves, likening executive leaders to patients, the role of the oversight group is to make accurate predictions, even if they differ, and the role of executive leadership is to determine the value those predictions have for the organization. For instance, an AI CDS system may have the potential to disproportionately operationally benefit one service line in terms of profits and efficiency while decreasing the revenue generated by another service line. The operational pros and cons could be presented objectively by the oversight group, and executive leadership would decide whether the level of harm would be acceptable in exchange for the predicted benefit. It is important to note here that organizational leaders can rightly determine the acceptable balance of organizational and operational benefit and harm, but *not* of patient benefit and harm. As will be addressed later in this chapter, it is essential that the acceptable level of patient harm associated with AI CDS be decided on, along with the relevant units of measurement, early on and be applied evenly as systems are considered.

5. **The oversight group should have a systematic process for evaluating each proposed system at each phase of the life cycle.** This chapter suggests a process to assess the *ethical* considerations as systems move from idea to clinical use. However, there are other important issues that should also be considered, such as workflow integration, legal implications, and risk management. The evaluation of each of these perspectives should be consistent, transparent, and objective. The evaluation should also consider concerns relevant to each of the stakeholder groups.

In addition, there should be some specific considerations applicable to systems that are purchased from a vendor. Vendors may not be completely transparent in sharing information about a system's limitations.[8] Because there is currently no regulatory requirement for how a system must be validated or what disclosures are necessary, it is incumbent upon the health system purchasing the AI system to independently validate the AI CDS. This is needed to establish both that the AI CDS performs as the vendor claims and that the AI CDS performs accurately in the local environment. Once the system has been validated in the local environment, even if its clinical benefit has been previously proven, it is important to prospectively study the AI system in the setting where it will function.[9] This is a kind of ethical insurance that the AI CDS is conferring a superior clinical benefit over the physician acting alone.

6. **The oversight group should be self-reflective.** Self-reflection is implicitly humble if it recognizes that each individual has the capacity to continue to improve and evolve and that identifying areas for improvement is the first step in that process. As such, there should be a core set of competencies expected of each member of the oversight group.[10] No one member can have possibly mastered all of the expertise that the group collectively possesses. In addition, the group should periodically self-evaluate in a structured and transparent manner. Some questions for reflection include the following:

   Are we multidisciplinary?

   Are we diverse?

   Are we/am I competent? How do we/I know? How do we/I measure our/my competency?

   Do we have an effective communications structure within our group and within the organization?

   When concerns about a system are raised, how do we/I address them? How do we/I know that it is an effective approach?

   Do we have the right people in the oversight group? Who else do we need? Am I still needed? Is some other member of the group still needed? What justifies continued membership in the group?

7. **There should be an agreed on process for sharing information about AI CDS systems extramurally.** Whether systems are developed internally or purchased from a vendor, it is important for health systems and payers to share important findings related to their validation, proof of clinical benefit, and/or deployment that might affect stakeholder groups, especially if there is the potential for harm. Since regulation and oversight of AI CDS is at this point still in flux and there is no formal structure for reporting safety concerns, organizations may hesitate to share information.[11] Nonetheless, health systems and payers have an obligation to alert other organizations considering or already utilizing AI CDS when there is the potential for harm. While harm might be viewed narrowly as bodily injury, such a definition would miss opportunities to prevent other important types of harm such as might arise if a commonly used training dataset is found to function less well in making clinical predictions when applied to a specific group of people. Such an instance would be harmful even if the system's predictions did not result in bodily injury.

While sharing information that would prevent harm from an AI system is of particular concern, it is just as important that information that proves the clinical superiority of an AI system also be shared so that others can enjoy the benefit.[12] In contrast to an AI system that might improve hospital and payer operations, AI CDS systems affect patients and their medical care directly. Because of the application of these systems to patients, prospective trial results that prove a clear benefit to patients should be shared so that other patients can also benefit. There are AI systems that may have an operational aim, such as reducing the length of hospital stays, but are also found to confer a clinical benefit to patients, such as reducing the interval between arriving at the hospital and initiating treatment. In such a case, although the system has an operational goal, the clinical benefit conferred requires that the information be shared so that other patients can also benefit.

This points to the need for AI CDS to be subject to prospective clinical trials, not just to establish its benefit for use in a particular health system, but also to report the outcomes more broadly so that others might also benefit from the findings.[13] The sharing of beneficial clinical findings is a basic ethical obligation of medicine and medical research. The reporting of outcomes of clinical research on drugs and devices may be incentivized by prestige to the investigators and/or profit to the manufacturers. Equivalent prestige and profit are yet to be established in the area of AI CDS, and there may even be a disincentive to report prospective study outcomes because of the lack of regulation and concerns about liability.[14] There may even be a disincentive to report the benefit of a system so as to maintain an advantage over competing health systems or payers. Although currently there is not a robust regulatory environment, one can imagine that failure to share outcomes data that would be beneficial to patients, based on greed, would not be looked on favorably.

8. **Harm and benefit should be established by the oversight group.** This entails (1) defining harm and ben-

efit such that the terms can be applied to each system, (2) remaining focused on harm and benefit from the patient's perspective so as to avoid conflicts of interest, (3) establishing acceptable and unacceptable levels of harm and benefit, (4) identifying specific harms and benefits that can be audited consistently across systems, and (5) being alert for unintended and unforeseen harms and benefits. This discussion intentionally addresses harm and benefit, rather than safety and efficacy, in order to emphasize the perspective of the patient. While thresholds for safety and effectiveness may be easier to objectively establish, harm and benefit relate more to the lived experience of the patient. Safe and effective are necessary but not sufficient requirements for AI systems that are being used to support clinical decisions. These systems must demonstrate a clinical benefit beyond performing as they are programmed to do.

A. *Establishing harm and benefit*: Harm and benefit should be established in terms that facilitate their identification and measurement.[15] For instance, harm can generally be defined as injury, and benefit as the achievement of a desirable outcome. For example, take an AI CDS system that detects early stage, preinvasive breast cancer on mammography before the lesion is apparent to a trained radiologist. Benefit in this setting is the earlier detection of breast cancer, leading to potentially less radical surgery, a shorter recovery, and a better prognosis. If the system performs poorly on Asian women compared to other ethnic groups, the Asian women whose mammograms had been evaluated by the system may have had their cancers detected at a later stage and with a less favorable prognosis than the other women. The harm that the group of Asian women suffered was potentially more extensive breast surgery, longer recovery, and even a poorer prognosis.

While potential harms and benefits can be identified at each stage of the life cycle, it will only be during validation and when studying these systems prospectively that actual outcomes can be assessed and their association with the AI system established. Many harms can be reliably predicted, and benefits anticipated, prior to the application of AI CDS to actual patients. This provides an opportunity to both try to resolve potential harms prior to validation and also be alert for harms after deployment. The anticipated benefit of the system will likely be one of the first features to be defined. During the validation phase both the achievement of benefit and the extent of benefit will be determined.

B. *Avoiding conflicts of interest:* An ethically relevant situation that deserves consideration is when the same event causes harm to one stakeholder group and benefit to another.[16] When one of those two stakeholder groups is patients, then a conflict of interest might develop. For example, take an AI CDS system that predicts which patients have occult chronic kidney disease and alerts their physician. The physician is then instructed to order additional tests; verify the prediction; inform the patient if the prediction is verified; and write a note in the patient's EMR detailing the evaluation, diagnosis, and decision whether or not to begin medication to halt the progression of kidney disease. The patient may view as a benefit the fact that their kidney disease was detected early and that progression was halted. In contrast, the health system might view the AI CDS as a harm because it increased the length of an office visit, as the AI CDS required additional conversations, orders, and documentation. The health system's physicians were able to see fewer patients in a day, resulting in a loss of revenue. In this case, the chronic kidney disease AI CDS was beneficial to patients by halting the progression of their disease but potentially harmful to the health system because it resulted in the loss of revenue. In this case, there would be a conflict of interest between the patient's benefit and the health system's profitability.

C. *Establishing acceptable levels of harm and benefit*: If the importance of describing harm and benefit is so that they can each be identified, then the importance of measuring harm and benefit is so that they can be quantified, and acceptable and unacceptable levels can be defined. How are harm and benefit measured? Patients are likely to define benefit in terms of their desired outcome and measure it in terms of the number of life-years prolonged, the level of pain reduced, or the amount of function improved. It is important for an oversight group to quantify the risks and benefits of an AI system so that thresholds can be set to decide whether to continue the development and deployment of a system, to halt or abandon the project, or to decommission a functioning system.[17]

In general, three thresholds should be decided upon: (1) what the absolute greatest level of tolerable harm is, (2) what the absolute minimum level of tolerable benefit is, and (3) what the highest ratio of harm to benefit that would be tolerable is. In quantifying harm and benefit, it is essential to keep in mind that the patient is central to the entire endeavor. It is to maximize the patient's benefit and to minimize the patient's harm that AI CDS, payers, and health systems exist. Regardless of their individual interests in maximizing their own benefit and minimizing their own harm, their primary interest must be to increase the patient's benefit and decrease the patient's harm.

One way of thinking about the acceptable level of benefit and harm derived from AI CDS is to compare the acceptable levels of harm and benefit from humans compared to AI CDS when performing the same function. For instance, when a human oncologist prescribes a chemotherapy regimen for patients, what is considered an acceptable level of benefit to patients, an acceptable level of harm to patients, and an acceptable ratio of harm to benefit?[18] Using acceptable human outcomes as a guide, the oversight group can decide whether the AI CDS should be held to the same standard or a higher standard than humans.[19] At times there will also be multiple possible harms and benefits that patients could experience. In this case it is important to be as quantitative as possible in considering what benefit is most beneficial and what harm is most harmful. One challenge is that different patients are likely to place different values on harms and benefits. This variation could be mitigated by considering what the intended benefit of the AI CDS is and what proximate harm could occur as a result of AI CDS being introduced into the relationship.

D. *Identifying specific harms and benefits*: In addition to considering system-specific direct patient benefits and harms, it is also important to think about broader known harms that are associated with AI CDS generally, such as equity, bias, and inconsistent system performance across populations.[20] The potential for these kinds of harms will be present regardless of what specific prediction the system is making or what specific patient the system is making a prediction for. Knowing that these kinds of harm can occur, and in what ways, allows the oversight group to audit for them and potentially address and resolve them at various stages in the AI CDS life cycle.

E. *Unintended and unforeseen harms and benefits*: In addition to predicting benefit and harm and looking for known harms and benefits, there is also the possibility that AI CDS will be associated with unintended and unpredicted harms and benefits. Both in terms of harm and benefit, it is important for the oversight group to be aware of this possibility, in particular being alert for safety signals that could mean ongoing harms and the need to halt or decommission the system in order to avoid further, ongoing harm. This comprehensive approach to benefit and harm not only serves to protect the patient and establish the clinical utility of the AI CDS but also builds trust in the AI CDS among physicians, patients, health systems, and payers.[21] It is plausible that a system that has been well designed and well trained, and that confers much greater benefit than harm, would be underutilized because clinicians and patients do not trust it. Having an established and trustworthy process for overseeing AI CDS will lead to greater adoption and utilization of the system.

ETHICAL CONCERNS IN THE AI CDS LIFE CYCLE

Exploratory Phase

During this phase, the clinical problem is formulated that the AI system is intended to address. The system at this point is theoretical; however, important ethical concerns arise that should be addressed before moving forward with the project.

A. Is the problem being addressed because it is clinically significant, and will addressing it be beneficial to patients? Is there true clinical equipoise around this problem? Although there may be other benefits to addressing the proposed issue, such as increased operational efficiency that may result in increased revenue, it is essential that the primary goal be to improve the clinical care of patients.[22] It is also at this stage that potential conflicts of interest that might be relevant throughout the AI system's life cycle are first identified.

B. How will benefit and harm be defined, measured, and proven? Similarly, it is important to think about how benefit and harm will be distributed. Are certain groups more likely to benefit? Will the benefit accorded to one group translate into harm to another group.[23] For instance, imagine an AI CDS system that confers so much clinical benefit that anyone who doesn't have access to it can not possibly achieve the same level of benefit. The group without access would be harmed because they do not have access to the same level of benefit.

C. Are there equity concerns? Similar to questions about the distribution of benefit and harm in the system's performance, one might ask whether the system is addressing an issue that is relevant across groups.[24]

Development and Training Phase

After a clinically significant issue has been decided on, with measurable and equitably distributed benefits and harms, the next phase involves programming and training the system to make the desired prediction, validating its accuracy, and prospectively establishing its clinical benefit.

A. Training: Ethical concerns raised in the training phase largely relate to the data being used to train the system. The system will only perform as well as the data on which it is trained, and biased training data will teach the system to make biased predictions.[25]
   a. Are the available training data diverse?
   b. Are some groups underrepresented or overrepresented?
   c. Do the data represent best clinical practices, or is there the potential for the system to learn from historically biased practices?
B. Validation: The validation phase is the first time that the system is tested on actual patients. The main goal is to

establish that the system accurately makes the prediction that it was trained to make. During this phase harms and associated trends can also be identified.[26] For commercially purchased systems, it is essential that they be validated in the local environment because of the potential for dataset shift.[27] Careful analysis of data from the validation phase will also identify any equity concerns. In addition to specifically focused questions, part of the debriefing after the validation phase is completed should ask broader questions about what was learned about the system, particularly what issues could be addressed through greater stakeholder involvement, education, and information.

a. What is being learned about the system as it is validated?
b. How are patients being informed about the use of an AI system? What are their concerns? What additional information/education is needed?
c. Are physicians utilizing the system? What are their concerns? What additional information/education is needed?
d. Are there gaps in the system's performance?
e. What groups of patients were chosen for validation?
f. Are some groups benefited and/or harmed more or less than other groups?
g. Has the system met the threshold for benefit?
h. Has the system exceeded the threshold for harm?
i. What adjustments need to be made to the system?
j. Should the system move on to a prospective study to prove clinical benefit?
k. If the decision is made to move forward, what kind of patient and family, provider, and health system edu-

cation is needed in order to build trust and promote use of the system?

C. Prospective study proving the clinical benefit of the system: Whereas the validation phase establishes the performance of the system, the system's clinical benefit has not yet been established. This is achieved, or not, by prospectively comparing the system to standard practice. The patients on whom the system is studied are research subjects, and in many institutions these studies will be conducted under the auspices of the IRB.[28] That adds ethical considerations regarding the patient not only as a patient but also as a research subject.
   a. How is benefit being defined and quantified?
   b. What is the end point of the study? How will it be known whether or not clinical benefit has been proven?
   c. What is the mechanism for identifying harm? Is there an auditing structure? Is there a structure for reporting safety as well as ethical concerns? How will it be known that the study should be stopped because of safety concerns?
   d. What is being done to educate/inform physicians, other clinicians, and any other staff who might come into contact with the AI CDS system, even indirectly?
   e. What is being done to educate/inform patients and families who are selected to be part of the study? How will patients in the control arm be educated/informed? Patients in the AI system arm? What does the consent process look like? Has the IRB given approval?

f. Once the study is complete, if clinical benefit is proven, should patients in the control arm be given access to the system if they would benefit from it?

g. What additional information was learned from the study?

The Deployment Phase

Deploying AI CDS in clinical practice is in some ways a pinnacle achievement in the system's life cycle, and in other ways it is just the beginning of a much longer process of learning and auditing. The ethically relevant concerns that arise relate to preparing physicians, patients and families, and other clinical staff for when the system goes live.

A. Physician Education and Information: A large portion of the information provided to physicians will be operational and will address how to integrate AI CDS into their workflow. However, there are some very specific ethical concerns that should be addressed at this stage in order to avoid confusion, over- or under-usage of the systems, or potentially substandard medical care.[29] Not only will education decrease the risk of harm from using the AI system, but it will also serve to build trust between physicians and AI systems.[30]

   a. How will physicians be educated about core AI competencies?

   b. How will physicians be educated about the specific AI CDS being deployed, including the limitations of the AI system?

   i. One way of presenting this information is to have a label similar to that used for nutritional information. Such labels might include the following information: a description of the AI CDS, its intended use, known uses, restrictions on use, when not to use, and specific information about the dataset used for training.[31]

c. Who will provide physician education, and where can physicians get additional information in real time?

d. How will overdependence be avoided?

   i. While health systems will rightly want to promote adoption of clinically beneficial AI systems by physicians, it is also important for physicians to recognize that these are predictions, not diagnoses, and that the prediction should be evaluated in the context of clinical judgment. An AI system that predicts with a high degree of accuracy whether a patient has an occult colorectal cancer will still require a colonoscopy, biopsy, and histological confirmation before beginning treatment for the cancer. Clinicians should not be overly influenced by automation bias and abdicate their clinical judgment.[32]

   ii. There is also concern that certain reimbursement models may promote overdependence on AI CDS. In particular, per-use reimbursement may incentive the use of AI CDS and create a conflict of interest on the part of physicians.[33]

e. How will disagreements between the AI CDS and physicians be resolved?

i. Presuming that upon deployment the AI CDS has not been adopted as the "standard of care," it will be important to help physicians know how to resolve disagreements that they will inevitably have with the AI system. There should be a process in place to help physicians resolve conflict so as to both encourage use and also avoid relinquishing clinical judgment. Disagreements have the potential to leave physicians uncertain about how to proceed and may actually lead to more moral distress than had they not collaborated with the AI system.[34]

f. How will physicians explain the AI CDS prediction to patients? How will physicians inform patients that they collaborated with AI CDS in making a clinical decision? Should patients be informed if the physician rejected the AI system's prediction?

g. Will physicians be able to "opt out" of using AI CDS?

i. Presuming that the clinical benefit of the AI system being deployed has been proven, it would be ethically problematic to allow physicians to not utilize a tool that has been shown to benefit patients. It would be similar to allowing physicians to opt out of using certain medications that have been proven beneficial or certain diagnostic tests. Every effort should be made to understand the physician's reasoning and to engage in a meaningful way with the physician. Ultimately, if the AI CDS has been proven to be clinically beneficial to patients, physicians have an obligation to utilize the system unless their patient refuses.

B. Patient and Family Education and Information: For patients and families, the central ethical theme in preparing to deploy AI CDS is how best to inform patients about the AI system and the decision that it is supporting, and how to provide an equivalent clinical pathway if they refuse the use of AI CDS in their medical decisions. Thinking of the patient as a patient, as a data donor, and as a learning subject, it will be essential to inform patients about all the roles. Patients' concerns about the AI system being used may be addressed effectively by evidence from a prospective study proving its clinical benefit. On the other hand, patients may have lingering concerns about privacy and the use of their data that dissuade them from allowing collaboration with AI CDS.

   a. How will patients and families be educated about AI?
   b. How will patients and families be educated about the specific AI CDS being deployed, including the benefits and limitations of the AI system?
   c. How will the AI CDS prediction be explained to patients?
      i. It should be noted that there is an important difference between informing patients that AI CDS is being used to make a clinical decision and explaining that decision to them. Explainability should be addressed as part of AI education generally and when it comes to the specific AI CDS system being used. Again, prospective studies proving the clinical benefit of AI CDS will help build trust in the system's precision. In the absence of an explanation of the reasoning behind a prediction, trust in the system's prediction based on proof of clinical

benefit will be very important. Additionally, trust will be built when the patient is informed and educated about AI generally and the specific AI CDS being used, including addressing data and privacy concerns.

d. Who will educate the patient about AI and explain the AI CDS system to the patient? Where can patients and families get additional information?
   i. Physicians may be ill equipped to fill this role entirely. While they have the ethical onus for making sure that the patient has been informed, it is certainly desirable for there to be standardized communications and perhaps expert resources available to patients if they request them.[35]

e. How will informed refusal be handled? What equivalent clinical processes will be in place?

C. Staff Education and Information: As important members of the patient's care team, all clinical and nonclinical staff would benefit from general education about AI and AI CDS. Many times patients will share their concerns and questions with nonphysician clinicians more readily than they will with physicians. Thus, it is important for all members of the care team to be informed. Trust in AI and AI CDS can be achieved in large part through education and engagement.[36]

a. How will staff members be educated about AI?

b. How will staff members be educated about the specific AI CDS being deployed, including the benefits and limitations of the AI system?

c. Who will provide staff education, and where can staff members get additional information in real time?

Ethical Concerns in the Clinical Use of AI CDS

Whereas validation establishes that AI CDS is accurate, and a prospective study establishes that it is beneficial, it is after deployment that the system has the greatest potential to do harm. Thus, there must be mechanisms in place for auditing for harm and for reporting concerns. As harm will inevitably occur, it is important to establish the threshold for pausing or decommissioning the system, how to learn from the events, and how to assign accountability when harm occurs.

A. Auditing for and Reporting Harm: Before systems are deployed, an auditing mechanism should be established. This would include the frequency of audits, what will be audited for, who will perform the audits, and how the audits will be reported. Auditing for harm is essential throughout the life cycle of the deployed AI CDS, especially early on, after system updates, and if there were events that could predictably lead to dataset drifts such that the accuracy of the system decreases. In addition, there should be a system in place whereby safety and ethical concerns could be reported, particularly concerns that need immediate attention in order to prevent further harm.[37]

   a. How frequently will the system be audited, and by whom?
   b. What are the markers that may predict dataset drift, and will they trigger more frequent audits?
   c. How often will the system be updated? When after an update will an audit occur?
   d. How often will the system be revalidated?

e. What might predict the need to reestablish the system's clinical benefit?
f. How will the audit results be communicated? Who is responsible for reviewing and acting on the results?
g. Will there be regular debriefing sessions after audits that involve the entire oversight group?
h. How will equity, bias, and other systemic harms be audited?
i. How will unforeseen and unintended harms be audited?
j. How will patient satisfaction be audited?
k. How will physician and staff satisfaction be audited?
l. How will unintended uses be audited?
m. What is the reporting system for safety concerns? What is the reporting system for ethics concerns? What is the reporting system for escalating concerns that need immediate attention?

B. Thresholds for Pausing or Decommissioning the AI CDS: There will be some harms that will not be apparent until the AI CDS has been used to support decisions for many more patients than in the validation and prospective study phases. The threshold for pausing and/or decommissioning the system should be decided well before the system is deployed. If a system is paused because of concern for ongoing harm, and the source of harm is addressed, there should be a period of close and careful observation to assure not only that the harm in question was adequately resolved but that no other harm has arisen.[38]

a. What is the established definition of harm relative to the AI CDS, how is it measured, and what is the threshold for pausing/decommissioning the system?
b. Who is responsible for the final decision to decommission?
c. If a system is paused in order to address a concern about harm, how will the system be audited once it is redeployed?
d. How will an AI CDS system be debriefed after it is paused or decommissioned?

C. Assigning Accountability for Harm: Assigning responsibility for harm due to AI systems generally, and to AI CDS specifically, is a source of debate. While many solutions are proposed, no single approach has been widely accepted in practice. It is generally thought that responsibility should not be assigned to the physician alone, but more appropriately should be distributed among multiple stakeholders.[39] One intriguing proposal is for AI insurance that would help foster adoption of AI systems by indemnifying the user and shifting part of the onus to underwriters for assuring that AI CDS systems are safe, effective, and support evidence-driven health care.[40]

It is also important to decide in advance who will disclose harm to patients and families and who will address the emotions of physicians and staff involved in the patient's care.

a. How will a root cause analysis (RCA) or debrief of harm due to AI CDS be conducted? Who is responsible for conducting the RCA or debrief?

b. What is the role in the RCA or debrief of the hospital quality and risk management representatives to the oversight committee?
c. How will it be determined if reporting the harm to any licensing or regulatory agency is required? Who is responsible for the report?
d. When harms are identified, how will the harm be disclosed to the affected patients and families? Who will disclose the harm?
e. When harms are identified, how will distress on the part of physicians and clinical and nonclinical staff be addressed? Whose role is it to do so?
f. How will it be decided whether the patient should be compensated for the harm? How will the extent of compensation be determined?

CASES

Case Study 6.1: AI-Assisted Robotic Surgery

Dr. Stark is an early adopter of robotic-assisted surgery platforms, in which a surgeon at a controller console uses finger clutch sensors to manipulate the instrument arms of a surgical robot hovering over the patient. Early robotic surgery platforms give surgeons high-definition, three-dimensional operative field views. Subsequent platforms include an additional safety feature: computer vision–aided boundary detection and avoidance of critical structures. A real-time computer vision algorithm uses pixels as model inputs to identify blood vessels that shouldn't be touched during surgery and links to the robotic instrument arms to prevent them from touching—and potentially injuring—those vessels.

Mr. Addison is a healthy, physically active seventy-one-year-old man who suffers a high-speed motor vehicle crash, fracturing two ribs and sustaining a concussion, but recovers quickly. A computed tomography scan of his abdomen, performed to look for internal bleeding, identifies a 5 cm right adrenal gland tumor that is probably benign but has some potentially malignant characteristics. He is referred to Dr. Stark, who recommends robotic right adrenalectomy and conveys to the patient that the major risk of this surgery is injury to the inferior vena cava—the largest vein in the body, immediately adjacent to the right adrenal gland. Dr. Stark explains that he has never injured the IVC during adrenalectomy and will use a new robotic surgery safety feature that identifies and protects the IVC during surgery. The benefit of surgery is tumor removal, allowing definitive determination of malignancy; if it is malignant, surgery may be curative. Mr. Addison plans to undergo surgery a full month before traveling to his niece's wedding, allowing adequate time for postoperative rest and recovery.

During surgery, Dr. Stark makes swift progress in identifying and exposing the right adrenal gland, which is soft and buttery, probably containing a lipoma (benign fatty tumor). He tests the boundary detection software by gently moving an instrument toward the IVC and is gratified to note that the robot senses danger and freezes the instrument before it makes contact, but the robot does allow him to touch the adrenal vein, which must be divided to remove the gland. He dissects the groove between the adrenal gland and IVC, looking for the small adrenal vein between them and preparing to close both ends with metallic clips before dividing it. Dr. Stark uses an instrument to push the gland away from the IVC to create a landing zone for the clips.

He pushes too hard. The adrenal vein tears at its confluence with the IVC. The controller console screen and wall-mounted screens go dark as blood sprays the robotic camera lens. A surgical technician swiftly withdraws the camera from inside the abdomen, cleans the blood off the lens, and carefully reinserts the camera on a new trajectory away from the arc of blood. As Mr. Anderson's blood pressure falls and heart rate rises, Dr. Stark feels his own heart rate rise, but remains calm. He takes a deep breath, exhales slowly, moves his instruments into position to control the bleeding and repair the IVC, and—nothing. His instruments freeze millimeters away from the IVC, halted by the boundary detection software.

Dr. Stark calls for an emergency undocking of the robot while he prepares to convert to an open surgical approach (i.e., hands inside the abdomen). A circulating nurse retrieves and opens a new tray of instruments for open surgery. Dr. Stark makes a long incision just beneath the ribs and evacuates a large volume of shed blood, which continues to flood the operative field until Dr. Lee, one of his partners, arrives and helps him control and repair the injury while the anesthesiologist transfuses several units of red blood cells and other blood products. Mr. Addison stabilizes. The tumor is removed and sent to the pathology department. Dr. Lee can see that her partner is shaken and gently offers to close while he speaks with Mr. Addison's family. They appreciate his candor and remain hopeful that Mr. Addison will still recover in time for his niece's wedding.

Exceeding expectations, Mr. Stark makes a speedy recovery. Despite his stoicism, he grimaces at the sharp pains under his ribs while getting in and out of bed for daily laps around the hospital ward. Days later, the pain is subsiding and is controlled by oral pain medications. Mr. Addison is discharged home in good condition four days after surgery.

Three days later, Mr. Addison develops fever and worsening abdominal pain. In the emergency department, a computed tomography scan of the abdomen confirms that Mr. Addison has developed an intra-abdominal abscess—a pocket of infected fluid. The body resorbs sterile fluid after surgery, even when the residual fluid is blood that contains iron—food for bacteria—because the immune system can usually fight off small amounts of bacteria. Unfortunately, Mr. Addison's immune system was suppressed by blood transfusions during surgery. Additional testing shows that the infection has spread into Mr. Addison's bloodstream. Antibiotics can treat the bloodstream infection, but not the large abscess. Dr. Stark consults with an interventional radiologist, who agrees to place a drain through the abdominal wall into the abscess, guided by computed tomography imaging. This approach is preferable to draining the abscess by performing another surgery, which would be unnecessarily invasive and would risk injury to the intestines, which tend to be swollen and friable one week after open abdominal surgery. Unfortunately, the intestines are also vulnerable to injury during drain placement. Mr. Addison's condition goes from bad to worse when he incurs a bowel injury. The intestines can no longer be used safely to maintain nutrition, but the other option—parenteral nutrition delivered through an intravenous catheter—allows the bloodstream infection to persist.

The pathology report is finalized, confirming that the tumor was benign. The good news is bittersweet against the realizations that Mr. Addison will be unable to travel to his niece's wedding and that his prolonged critical illness will continue to threaten his health and quality of life long after discharge from the ICU and hospital.

QUESTIONS

Does Dr. Stark, who decided to use the computer vision–aided protection feature, bear sole responsibility for his inability to quickly gain control of the bleeding?

This scenario may have been prevented with a manual human override, but overrides themselves are associated with risk, as evident in the Chernobyl nuclear accident. Should manual overrides be included in surgical robotic guidance platforms?

Case Study 6.2: Human-Machine Conflict

Dr. Edwards recommends that her patient, Mr. Jones, be taken to the operating room to have a laparoscopic cholecystectomy (gallbladder removal performed through four small incisions using a camera to visualize the operation). To remove the gallbladder, the tube connecting the gallbladder to the main bile duct, which is called the cystic duct, must be correctly identified and cut. Surrounding tubes, such as the main bile duct and the right hepatic duct, look very similar to the cystic duct and must be differentiated from the cystic duct to avoid cutting the wrong structure, resulting in the uncontrolled leakage of bile into the abdomen, lack of bile drainage from the liver, or both.

The incorrect tube is cut in approximately 0.3 percent of the laparoscopic cholecystectomies performed, resulting in substantial patient suffering and lawsuits against surgeons and hospitals. To reduce the number of times the incorrect tube is identified and cut, Dr. Edwards's hospital, Community Hospital, has purchased a video-assisted AI CDS system that has been trained to accurately identify the correct duct, the cystic duct, so that the surgeon can cut it. Dr. Edwards uses this AI CDS system in Mr. Edwards's surgery.

When it is time to cut the cystic duct and remove the gallbladder, the system identifies a tube on the right side of the screen as the cystic duct. Dr. Edwards identifies a different tube, in the center of the screen, as the cystic duct. Dr. Edwards injects dye into the duct and takes an X-ray to better identify the correct duct to cut, but the X-ray does not help her to identify the correct duct.

QUESTIONS

What should Dr. Edwards do?

Option A: Dr. Edwards divides the tube that she believes to be correct. She is correct; the system was incorrect in its prediction. Had Dr. Edwards followed the system's guidance, the patient would have been harmed.

What should Dr. Edwards do next? Was there any harm done?

Option B: Dr. Edwards divides the tube that the system predicts to be the correct tube, rather than the tube that she believes to be correct. The system was correct in its prediction. Had Dr. Edwards ignored the system's guidance, the patient would have been harmed.

What should Dr. Edwards do next? Was there any harm done?

Should the patient have been informed that Dr. Edwards would be sharing critical intraoperative decisions with the AI CDS system? What harms could arise if the patient refused to allow the AI system to participate in his operation?

Case Study 6.3: Dataset Drift

America's Greatest Hospital (AGH) develops and implements a state-of-the-art AI algorithm that uses data collected within EMRs during standard clinical care to predict the development of bacterial sepsis, a time-sensitive condition for which delays in treatment are associated with increased risk of death. The algorithm performs well on retrospective (collected in the past) data and performs nearly as well on prospective (real-time) data. The algorithm's developers are confident that the algorithm is optimized. AGH physicians are confident that the algorithm's predictions will aid in the early recognition of bacterial sepsis. AGH administrators agree to implement the algorithm within EMRs under the watchful eyes of a data safety and monitoring board.

For two years, the algorithm works well. The interval between the arrival of a septic patient in the emergency department and initiation of sepsis therapy decreases, as does hospital mortality. Those metrics fluctuate as AGH and the rest of the world are wracked by a viral pneumonia pandemic. As the dust settles, the data safety and monitoring board notes that in addition to the observed fluctuations in time to sepsis treatment initiation and hospital mortality, the predictive performance of the bacterial sepsis prediction model is degrading. Further investigation demonstrates that the viral pneumonia pandemic has altered the previously observed association between fever and bacterial sepsis. The algorithm is decommissioned while developers seek deeper understanding of the dataset drift problem and test solutions.

QUESTIONS

AI algorithms can be locked (finished learning and capable only of making predictions based on what has been learned from training data in the past) or online (continuing to learn from new data as they are received). What are the bioethical trade-offs between locked and online learning?

Of the many involved stakeholders, who bears ultimate responsibility for the AI system's performance? How can that stakeholder establish accountability by other stakeholders?

NOTES

1. INTRODUCTION TO AI CLINICAL DECISION SUPPORT SYSTEMS

1. Sharon E. Kessler, "Why Care: Complex Evolutionary History of Human Healthcare Networks," *Frontiers in Psychology* 11 (2020): 199.

2. *Physician* is used in this text to refer to practitioners of medicine, the agents making clinical decisions, and in turn engaging with patients in shared decision-making. This is not intended to ignore the important contributions that nonphysician clinicians make to patient care and their rightful role in diagnosing and treating disease and offering comfort to patients. The focus on physicians is largely based on the distinct ethical obligations physicians owe to their patients within the larger context of medical ethics. In addition, most nonphysician providers perform their duties under the sponsorship of a physician such that the physician, at least theoretically, would always play a role in making any medical decision. Finally, at least initially, it is likely that many of the decisions for which AI systems will offer support will be made primarily by physicians.

3. Carina Fourie, "Sufficiency of Capabilities, Social Equality, and Two-Tiered Health Care systems," In *What Is Enough? Sufficiency, Justice*

and Health, ed. Carina Fourie and Anette Rid (New York: Oxford University Press, 2017), 185–204.

4. Matthias Braun, Patrik Hummel, Susanne Beck, and Peter Dabrock, "Primer on an Ethics of AI-Based Decision Support Systems in the Clinic," *Journal of Medical Ethics* 47, no. 12 (2021): e3; Danton S. Char, Michael D. Abràmoff, and Chris Feudtner, "Identifying Ethical Considerations for Machine Learning Healthcare Applications," *American Journal of Bioethics* 20, no. 11 (2020): 7–17; Irene Y. Chen, Emma Pierson, Sherri Rose, Shalmali Joshi, Kadija Ferryman, and Marzyeh Ghassemi, "Ethical Machine Learning in Healthcare," *Annual Review of Biomedical Data Science* 4 (2021): 123–44; Luciano Floridi, Josh Cowls, Monica Beltrametti, Raja Chatila, Patrice Chazerand, Virginia Dignum, Christoph Luetge, et al., "An Ethical Framework for a Good AI Society: Opportunities, Risks, Principles, and Recommendations," In *Ethics, Governance, and Policies in Artificial Intelligence*, ed. Luciano Floridi, (Cham, Switzerland: Springer Nature, 2021), 19–39; Melissa D. McCradden, James A. Anderson, Elizabeth A. Stephenson, Erik Drysdale, Lauren Erdman, Anna Goldenberg, and Randi Zlotnik Shaul, "A Research Ethics Framework for the Clinical Translation of Healthcare Machine Learning," *American Journal of Bioethics* 22, no. 5 (2022): 8–22; Jessica Morley, Caio C. V. Machado, Christopher Burr, Josh Cowls, Indra Joshi, Mariarosaria Taddeo, and Luciano Floridi, "The Ethics of AI in Health Care: A Mapping Review," *Social Science & Medicine* 260 (2020): 113172; Sandeep Reddy, Wendy Rogers, Ville-Petteri Makinen, Enrico Coiera, Pieta Brown, Markus Wenzel, Eva Weicken, et al., "Evaluation Framework to Guide Implementation of AI Systems into Healthcare Settings," *BMJ Health & Care Informatics* 28, no. 1 (2021) : e10044; Sebastian Vollmer, Bilal A. Mateen, Gergo Bohner, Franz J. Király, Rayid Ghani, Pall Jonsson, Sarah Cumbers, et al. "Machine Learning and Artificial Intelligence Research for Patient Benefit: 20 Critical Questions on Transparency, Replicability, Ethics, and Effectiveness," BMJ 368 (2020); Jenna Wiens, Suchi Saria, Mark Sendak, Marzyeh Ghassemi, Vincent X. Liu, Finale Doshi-Velez, Kenneth Jung, et al. "Do No Harm: A Roadmap for Responsible Machine Learning for Health Care," *Nature Medicine* 25, no. 9 (2019):

1337–40; and Roberto V. Zicari, John Brodersen, James Brusseau, Boris Düdder, Timo Eichhorn, Todor Ivanov, Georgios Kararigas, et al., "Z-Inspection®: A Process to Assess Trustworthy AI," *IEEE Transactions on Technology and Society* 2, no. 2 (2021): 83–97.

5. Tom L. Beauchamp and James F. Childress, *Principles of Biomedical Ethics*, 8th ed. (New York: Oxford University Press, 2019).

6. McCradden et al., "Research Ethics Framework."

7. This text uses the terms *AI systems* and *AI CDS* interchangeably for simplicity, realizing that many of these systems utilize more complex processes such as machine learning and neural networks.

8. Bryan Pilkington and Charles Binkley, "Disproof of Concept: Resolving Ethical Dilemmas Using Algorithms," *American Journal of Bioethics* 22, no. 7 (2022): 81–83.

9. Basil Varkey, "Principles of Clinical Ethics and Their Application to Practice," *Medical Principles and Practice* 30, no. 1 (2021): 17–28.

10. K. Danner Clouser and Bernard Gert, "A Critique of Principlism," *Journal of Medicine and Philosophy* 15, no. 2 (1990): 219–36; and Luciano Floridi and Josh Cowls, "A Unified Framework of Five Principles for AI in Society," In *Machine Learning and the City: Applications in Architecture and Urban Design*, ed. Silvio Carta (Wiley Online Library, 2022), 535–45.

11. Ian Duncan, Tamim Ahmed, Henry Dove, and Terri L. Maxwell, "Medicare Cost at End of Life," *American Journal of Hospice and Palliative Medicine®* 36, no. 8 (2019): 705–10.

12. Charles E. Binkley, Joel Michael Reynolds, and Andrew Shuman, "From the Eyeball Test to the Algorithm—Quality of Life, Disability Status, and Clinical Decision Making in Surgery," *New England Journal of Medicine* 387, no. 14 (2022): 1325–28; and Charles E. Binkley, David S. Kemp, and Brandi Braud Scully, "Should We Rely on AI to Help Avoid Bias in Patient Selection for Major Surgery?," *AMA Journal of Ethics* 24, no. 8 (2022): 773–80.

13. Regina G. Russell, Laurie Lovett Novak, Mehool Patel, Kim V. Garvey, Kelly Jean Thomas Craig, Gretchen P. Jackson, Don Moore, and Bonnie M. Miller, "Competencies for the Use of Artificial Intelligence–Based Tools by Health Care Professionals," *Academic Medicine* 98, no. 3 (2023): 348–56.

2. THE PHYSICIAN AND AI CLINICAL DECISION SUPPORT SYSTEMS

1. Tom L. Beauchamp, and James F. Childress, *Principles of Biomedical Ethics*, 8th ed. (New York: Oxford University Press, 2019).

2. Thomas A. Cavanaugh, *Hippocrates' Oath and Asclepius' Snake: The Birth of the Medical Profession* (New York: Oxford University Press, 2018).

3. Stephanie A. Kraft, "Respect and Trustworthiness in the Patient-Provider-Machine Relationship: Applying a Relational Lens to Machine Learning Healthcare Applications," *American Journal of Bioethics* 20, no. 11 (2020): 51–53.

4. Cavanaugh, *Hippocrates' Oath*.

5. Stacy M. Carter, Wendy Rogers, Khin Than Win, Helen Frazer, Bernadette Richards, and Nehmat Houssami, "The Ethical, Legal and Social Implications of Using Artificial Intelligence Systems in Breast Cancer Care," *Breast* 49 (2020): 25–32.

6. Alex John London, "Artificial Intelligence and Black-Box Medical Decisions: Accuracy versus Explainability," *Hastings Center Report* 49, no. 1 (2019): 15–21; and Aaron Springer and Steve Whittaker, "Making Transparency Clear: The Dual Importance of Explainability and Auditability," in *Joint Proceedings of the ACM IUI 2019 Workshop, Los Angeles, CA, USA, March 20, 2019* (Association for Computing Machinery, 2019), https://ceur-ws.org/Vol-2327/IUI19WS-IUIATEC-5.pdf.

7. Matthias Braun, Patrik Hummel, Susanne Beck, and Peter Dabrock, "Primer on an Ethics of AI-Based Decision Support Systems in the Clinic," *Journal of Medical Ethics* 47, no. 12 (2021): e3; and Thomas Grote and Philipp Berens, "On the Ethics of Algorithmic Decision-Making in Healthcare," *Journal of Medical Ethics*, 46, no. 3 (2020): 205–11.

8. Braun et al., "Primer on an Ethics," e3.

9. Karl Y. Bilimoria, Yaoming Liu, Jennifer L. Paruch, Lynn Zhou, Thomas E. Kmiecik, Clifford Y. Ko, and Mark E. Cohen, "Development and Evaluation of the Universal ACS NSQIP Surgical Risk Calculator: A Decision Aid and Informed Consent Tool for Patients and Surgeons," *Journal of the American College of Surgeons* 217, no. 5 (2013): 833–42.

10. Braun et al., "Primer on an Ethics," e3; Carter et al., "Ethical, Legal and Social Implications," 25–32; and Grote and Berens, "Ethics of Algorithmic Decision-Making," 205–11.

11. Melissa D. McCradden, Elizabeth A. Stephenson, and James A. Anderson, "Clinical Research Underlies Ethical Integration of Healthcare Artificial Intelligence," *Nature Medicine* 26, no. 9 (2020): 1325–26; Melissa D. McCradden, James A. Anderson, and Randi Zlotnik Shaul, "Accountability in the Machine Learning Pipeline: The Critical Role of Research Ethics Oversight," *American Journal of Bioethics* 20, no. 11 (2020): 40–42; and Melissa D. McCradden, James A. Anderson, Elizabeth A. Stephenson, Erik Drysdale, Lauren Erdman, Anna Goldenberg, and Randi Zlotnik Shaul, "A Research Ethics Framework for the Clinical Translation of Healthcare Machine Learning," *American Journal of Bioethics* 22, no. 5 (2022): 8–22.

12. McCradden, Stephenson, and Anderson, "Research Underlies Ethical integration," 1325–26.

13. Derek C. Angus, "Randomized Clinical Trials of Artificial Intelligence," *JAMA* 323, no. 11 (2020): 1043–45; Dinah V. Parums, "Artificial Intelligence (AI) in Clinical Medicine and the 2020 CONSORT-AI Study Guidelines," *Medical Science Monitor: International Medical Journal of Experimental and Clinical Research* 27 (2021): e933675-1; and Eric J. Topol, "Welcoming New Guidelines for AI Clinical Research," *Nature Medicine* 26, no. 9 (2020): 1318–20.

14. McCradden et al., "Research Ethics Framework," 8–22.

15. Maia Jacobs, Melanie F. Pradier, Thomas H. McCoy Jr., Roy H. Perlis, Finale Doshi-Velez, and Krzysztof Z. Gajos, "How Machine-Learning Recommendations Influence Clinician Treatment Selections: The Example of Antidepressant Selection," *Translational Psychiatry* 11, no. 1 (2021): 108.

16. Stephanie Eaneff, Ziad Obermeyer, and Atul J. Butte, "The Case for Algorithmic Stewardship for Artificial Intelligence and Machine Learning Technologies," *JAMA* 324, no. 14 (2020): 1397–98.

17. Danton S. Char, Nigam H. Shah, and David Magnus, "Implementing Machine Learning in Health Care—Addressing Ethical

Challenges," *New England Journal of Medicine* 378, no. 11 (2018): 981; and Carter et al., "Ethical, Legal and Social Implications," 25–32.

18. Abeba Birhane, Pratyusha Kalluri, Dallas Card, William Agnew, Ravit Dotan, and Michelle Bao, "The Values Encoded in Machine Learning Research," in *FaccT '22: Proceedings of the 2022 ACM Conference on Fairness, Accountability, and Transparency* (Association for Computing Machinery, 2022), 173–84, https://dl.acm.org/doi/10.1145/3531146.3533083; and Carter et al., "Ethical, Legal and Social Implications," 25–32.

19. Carter et al., "Ethical, Legal and Social Implications," 25–32; Char, Shah, and Magnus, "Implementing Machine Learning," 981.

20. Jessica Morley, Caio C. V. Machado, Christopher Burr, Josh Cowls, Indra Joshi, Mariarosaria Taddeo, and Luciano Floridi, "The Ethics of AI in Health Care: A Mapping Review," *Social Science & Medicine* 260 (2020): 113172.

21. Angus, "Randomized Clinical Trials of artificial Intelligence," 1043–45; and Topol, "Welcoming new guidelines," 1318–20.

22. Jenna Wiens, Suchi Saria, Mark Sendak, Marzyeh Ghassemi, Vincent X. Liu, Finale Doshi-Velez, Kenneth Jung, et al., "Do No Harm: A Roadmap for Responsible Machine Learning for Health Care," *Nature Medicine* 25, no. 9 (2019): 1337–40.

23. Carter et al., "Ethical, Legal and Social Implications," 25–32; and Enrico Coiera, "The Fate of Medicine in the Time of AI," *Lancet* 392, no. 10162 (2018): 2331–32.

24. C. Gretton, "The Dangers of AI in Health Care: Risk Homeostasis and Automation Bias," *Towards Data Science* (blog), June 24, 2017, https://towardsdatascience.com/the-dangers-of-ai-in-health-care-risk-homeostasis-and-automation-bias-148477a9080f.

25. Himabindu Lakkaraju and Osbert Bastani, "'How Do I Fool You?' Manipulating User Trust via Misleading Black Box Explanations," in *AIES '20: Proceedings of the AAAI/ACM Conference on AI, Ethics, and Society* (February 2020), 79–85, https://dl.acm.org/doi/10.1145/3375627.3375833; and Marco Tulio Ribeiro, Sameer Singh, and Carlos Guestrin. "'Why Should I Trust You?' Explaining the Predictions of Any Classifier," in *Proceedings of the 22nd ACM SIGKDD International Conference on Knowledge Discovery and Data Mining* (Association for

Computing Machinery, 2016), 1135–44, https://www.kdd.org/kdd2016/papers/files/rfp0573-ribeiroA.pdf.

26. I. Glenn Cohen, "Informed Consent and Medical Artificial Intelligence: What to Tell the Patient?," *Georgetown Law Journal* 108 (2019): 1425; and W. Nicholson Price, Sara Gerke, and I. Glenn Cohen, "Potential Liability for Physicians Using Artificial Intelligence," *JAMA* 322, no. 18 (2019): 1765–66.

27. Cohen, "Informed Consent and Medical Artificial Intelligence," 1425; and Price, Gerke, and Cohen, "Potential Liability," 1765–66.

28. Price, Gerke, and Cohen, "Potential Liability," 1765–66; Philipp Hacker, Ralf Krestel, Stefan Grundmann, and Felix Naumann, "Explainable AI under Contract and Tort Law: Legal Incentives and Technical Challenges," *Artificial Intelligence and Law* 28 (2020): 415–39.

29. Julia Amann, Alessandro Blasimme, Effy Vayena, Dietmar Frey, and Vince I. Madai, "Explainability for Artificial Intelligence in Healthcare: A Multidisciplinary Perspective," *BMC Medical Informatics and Decision Making* 20, no. 1 (2020): 1–9; and Rosalind J. McDougall, "Computer Knows Best? The Need for Value-Flexibility in Medical AI," *Journal of Medical Ethics* 45, no. 3 (2019): 156–60.

30. Lukas J. Meier, Alice Hein, Klaus Diepold, and Alena Buyx, "Algorithms for Ethical Decision-Making in the Clinic: A Proof of Concept," *American Journal of Bioethics* 22, no. 7 (2022): 4–20.

31. Peter Moffett and Gregory Moore, "The Standard of Care: Legal History and Definitions; the Bad and Good News," *Western Journal of Emergency Medicine* 12, no. 1 (2011): 109.

32. Brian K. Cooke, Elizabeth Worsham, and Gary M. Reisfield, "The Elusive Standard of Care," *Journal of the American Academy of Psychiatry and the Law* 45, no. 3 (2017): 358–64.

33. Price, Gerke, and Cohen, "Potential Liability," 1765–66.

34. Price, Gerke, and Cohen, "Potential Liability" 1765–66.

35. Joel Michael Reynolds, Charles E. Binkley, and Andrew Shuman, "The Complex Relationship between Disability Discrimination and Frailty Scores," *American Journal of Bioethics* 21, no. 11 (2021): 74–76.

36. Bilimoria et al., "Development and Evaluation of Universal ACS NSQIP." 833–842.

37. Kim A. Eagle, Peter B. Berger, Hugh Calkins, Bernard R. Chaitman, Gordon A. Ewy, Kirsten E. Fleischmann, Lee A. Fleisher, et al., "ACC/AHA Guideline Update for Perioperative Cardiovascular Evaluation for Noncardiac Surgery—Executive Summary: A Report of the American College of Cardiology/American Heart Association Task Force on Practice Guidelines (Committee to Update the 1996 Guidelines on Perioperative Cardiovascular Evaluation for Noncardiac Surgery)," *Journal of the American College of Cardiology* 39, no. 3 (2002): 542–53.

38. Tyler J. Loftus, Patrick J. Tighe, Amanda C. Filiberto, Philip A. Efron, Scott C. Brakenridge, Alicia M. Mohr, Parisa Rashidi, et al., "Artificial Intelligence and Surgical Decision-Making," *JAMA Surgery* 155, no. 2 (2020): 148–58.

39. Beauchamp and Childress. "Principles of Biomedical Ethics"; and Bernard Lo, *Resolving Ethical Dilemmas: A Guide for Clinicians*, 6th ed. (Philiadelphia, PA: Wolters Kluwer) 2020.

40. Reshma Jagsi and Lisa Soleymani Lehmann, "The Ethics of Medical Education," *BMJ* 329, no. 7461 (2004): 332–34; and Richelle K. Marracino and Robert D. Orr, "Entitling the Student Doctor: Defining the Student's Role in Patient Care," *Journal of General Internal Medicine* 13 (1998): 266–70.

41. "NJ A1494," Bill Track 50, accessed September 8, 2023, https://www.billtrack50.com/BillDetail/1171771#:~:text=Bill%20Summary,consent%20to%20the%20invasive%20examination; and Maya M. Hammoud, Kayte Spector-Bagdady, Meg O'Reilly, Carol Major, and Laura Baecher-Lind, "Consent for the Pelvic Examination under Anesthesia by Medical Students: Recommendations by the Association of Professors of Gynecology and Obstetrics," *Obstetrics and Gynecology* 134, no. 6 (2019): 1303.

42. Cohen, "Informed Consent and Medical Artificial Intelligence," 1425.

43. Braun et al., "Primer on an Ethics," e3.

44. Braun et al., "Primer on an Ethics," e3.

45. "Surgical Risk Calculator," American College of Surgeons, accessed October 9, 2023, https://riskcalculator.facs.org/RiskCalculator/PatientInfo.jsp.

46. Charles E. Binkley, "How Should Surgeons Communicate about Palliative and Curative Intentions, Purposes, and Outcomes?," *AMA Journal of Ethics* 23, no. 10 (2021): 794–99.

47. Braun et al., "Primer on an Ethics," e3; and Grote and Berens, "Ethics of Algorithmic Decision-Making," 205–11.

48. Birhane et al., "Values in Machine Learning Research," 173–84; Carter et al., "Ethical, Legal and Social Implications," 25–32; and Char, Shah, and Magnus, "Implementing Machine Learning in Health Care," 981.

49. Morley et al., "Ethics of AI in Health Care," 113172.

50. Price, Gerke, and Cohen, "Potential Liability," 1765–66.

51. Price, Gerke, and Cohen, "Potential Liability" 1765–66.

52. Charles E. Binkley and Brian P. Green, "Does Intraoperative Artificial Intelligence Decision Support Pose Ethical Issues?," *JAMA Surgery* 156, no. 9 (2021): 809–10.

53. Braun et al., "Primer on an Ethics," e3; and Price, Gerke, and Cohen, "Potential Liability," 1765–66.

54. Cohen, "Informed Consent and Medical Artificial Intelligence," 1425; Price, Gerke, and Cohen, "Potential Liability," 1765–1766.

55. Braun et al., "Primer on an Ethics," e3; and Charles E. Binkley and Bryan Pilkington, "The Actionless Agent: An Account of Human-CAI Relationships," *American Journal of Bioethics* 23, no. 5 (2023): 25–27.

56. Binkley and Pilkington, "Actionless Agent," 25–27.

3. THE PATIENT AND AI CLINICAL DECISION SUPPORT SYSTEMS

1. Charles E. Binkley and Bryan Pilkington, "Informed Consent for Clinician-AI Collaboration and Patient Data Sharing: Substantive, Illusory, or Both?," *The American Journal of Bioethics* 23, no. 10 (2023): 83–85.

2. Carlos A. Pellegrini, "Trust: The Keystone of the patient-physician relationship." *Journal of the American College of Surgeons* 224, no. 2 (2017): 95–102.

3. National Commission for the Protection of Human Subjects of Biomedical and Behavioral Research, *The Belmont Report: Ethical Principles and Guidelines for the Protection of Human Research Subjects* (Office of the Secretary, Department of Health and Human Services, April 18, 1979), www.hhs.gov/ohrp/regulations-and-policy/belmont-report/read-the-belmont-report/index.html; Marc D. Basson, Gerald Dworkin, and Eric J. Cassell, "The 'Student Doctor' and a Wary Patient," *Hastings Center Report* 12, no. 1 (1982): 27–28; Daniel L. Cohen, Laurence B. McCullough, R. W. Kessel, Aristide Y. Apostolides, Errol R. Alden, and Kelly J. Heiderich, "Informed Consent Policies Governing Medical Students' Interactions with Patients," *Academic Medicine* 62, no. 10 (1987): 789–98; and Catharyn T. Liverman, and James F. Childress, eds., *Organ Donation: Opportunities for Action* (Washington, DC: National Academies Press, 2006).

4. Jennifer M. Barrow, Grace D. Brannan, and Paras B. Khandhar. "Research Ethics" In *StatPearls* (Treasure Island, FL: StatPearls Publishing; 2023), www.ncbi.nlm.nih.gov/books/NBK459281/.

5. Laura M. Beskow, "Lessons from HeLa Cells: The Ethics and Policy of Biospecimens," *Annual Review of Genomics and Human Genetics* 17 (2016): 395–417; Maureen S. Dorney, "Moore v. The Regents of the University of California: Balancing the Need for Biotechnology Innovation against the Right of Informed Consent," *High Technology Law Journal* 5 (1989): 333; and Saul Krugman, "The Willowbrook Hepatitis Studies Revisited: Ethical Aspects," *Reviews of Infectious Diseases* 8, no. 1 (1986): 157–62.

6. Ezekiel J. Emanuel and Linda L. Emanuel, "Four Models of the Physician-Patient Relationship," *JAMA* 267, no. 16 (1992): 2221–26.

7. Stacy S. Chen and Sunit Das, "'What Are My Options?': Physicians as Ontological Decision Architects in Surgical Informed Consent," *Bioethics* 36, no. 9 (2022): 936–39.

8. Jennifer Blumenthal-Barby, "An AI Bill of Rights: Implications for Health Care AI and Machine Learning—A Bioethics Lens," *American Journal of Bioethics* 23, no. 1 (2023): 4–6.

9. Nina Hallowell, Shirlene Badger, Aurelia Sauerbrei, Christoffer Nellåker, and Angeliki Kerasidou, "'I Don't Think People Are

Ready to Trust These Algorithms at Face Value': Trust and the Use of Machine Learning Algorithms in the Diagnosis of Rare Disease," *BMC Medical Ethics* 23, no. 1 (2022): 1–14.

10. Jordan P. Richardson, Cambray Smith, Susan Curtis, Sara Watson, Xuan Zhu, Barbara Barry, and Richard R. Sharp, "Patient Apprehensions about the Use of Artificial Intelligence in Healthcare," *NPJ Digital Medicine* 4, no. 1 (2021): 140.

11. Richardson et al., "Patient Apprehensions," 140.

12. Binkley and Pilkington, "Informed Consent."

13. Hippocrates. *Hippocratic Writings*, trans. G. E. R. Lloyd, John Chadwick, and W. N. Mann (New York: Penguin, 1978).

14. John Banja, "How Might Artificial Intelligence Applications Impact Risk Management?," *AMA Journal of Ethics* 22, no. 11 (2020): 945–51; and N. Nina. Zivanovic, "Medical Information as a Hot Commodity: The Need for Stronger Protection of Patient Health Information," *Intellectual Property Law Bulletin* 19 (2014): 183.

15. Banja, "How might applications impact risk management?," 945–51; and Brent Daniel Mittelstadt and Luciano Floridi, "The Ethics of Big Data: Current and Foreseeable Issues in Biomedical Contexts," *Science and Engineering Ethics* 22, no. 2 (2016): 303-41.

16. Jessica Morley, Caio C. V. Machado, Christopher Burr, Josh Cowls, Indra Joshi, Mariarosaria Taddeo, and Luciano Floridi, "The Ethics of AI in Health Care: A Mapping Review," *Social Science & Medicine* 260 (2020): 113172.

17. Robert Challen, Joshua Denny, Martin Pitt, Luke Gompels, Tom Edwards, and Krasimira Tsaneva-Atanasova, "Artificial Intelligence, Bias and Clinical Safety," *BMJ Quality & Safety* 28, no. 3 (2019): 231–37.

18. Richard F. Wagner Jr., Abel Torres, and Steven Proper, "Informed Consent and Informed Refusal," *Dermatologic Surgery* 21, no. 6 (1995): 555–59.

19. Stacy M. Carter, Wendy Rogers, Khin Than Win, Helen Frazer, Bernadette Richards, and Nehmat Houssami, "The Ethical, Legal and Social Implications of Using Artificial Intelligence Systems in Breast Cancer Care," *Breast* 49 (2020): 25–32.

20. Beskow, "Lessons from HeLa Cells," 395–417; and Dorney, "Moore v. Regents," 333.

21. Basson, Dworkin, and Cassell, "'Student Doctor' and Patient," 27–28.

22. Gary J. Becker, "Human Subjects Investigation: Timeless Lessons of Nuremberg and Tuskegee," *Journal of the American College of Radiology* 2, no. 3 (2005): 215–17.

23. Jason Dana and George Loewenstein, "A Social Science Perspective on Gifts to Physicians from Industry," *JAMA* 290, no. 2 (2003): 252–55.

24. Tom L. Beauchamp and James F. Childress, *Principles of Biomedical Ethics*, 8th ed. (New York: Oxford University Press, 2019).

25. Charles E. Binkley, David S. Kemp, and Brandi Braud Scully, "Should We Rely on AI to Help Avoid Bias in Patient Selection for Major Surgery?," *AMA Journal of Ethics* 24, no. 8 (2022): 773–80; Charles E. Binkley, Joel Michael Reynolds, and Andrew Shuman, "From the Eyeball Test to the Algorithm-Quality of Life, Disability Status, and Clinical Decision Making in Surgery," *New England Journal of Medicine* 387, no. 14 (2022): 1325–28; and Crystal N. Johnson-Mann, Tyler J. Loftus, and Azra Bihorac, "Equity and Artificial Intelligence in Surgical Care," *JAMA Surgery* 156, no. 6 (2021): 509–10.

26. Ziad Obermeyer, Brian Powers, Christine Vogeli, and Sendhil Mullainathan, "Dissecting Racial Bias in an Algorithm Used to Manage the Health of Populations," *Science* 366, no. 6464 (2019): 447–53.

27. Johnson-Mann, Loftus, and Bihorac, "Equity and Artificial Intelligence in Surgical Care," 509–10; Kristin M. Kostick-Quenet, I. Glenn Cohen, Sara Gerke, Bernard Lo, James Antaki, Faezah Movahedi, Hasna Njah, Lauren Schoen, Jerry E. Estep, and J. S. Blumenthal-Barby, "Mitigating Racial Bias in Machine Learning," *Journal of Law, Medicine & Ethics* 50, no. 1 (2022): 92–100; and Ravi B. Parikh, Stephanie Teeple, and Amol S. Navathe, "Addressing Bias in Artificial Intelligence in Health Care," *JAMA* 322, no. 24 (2019): 2377–78.

28. Parikh, Teeple, and Navathe, "Addressing Bias," 2377–78.

29. Binkley, Kemp, and Scully, "Should We Rely on AI," 773–80.

30. Kostick-Quenet et al., "Mitigating Racial Bias in Machine Learning." 92–100.

31. Bonnie Kaplan, "Selling Health Data: De-identification, Privacy, and Speech," *Cambridge Quarterly of Healthcare Ethics* 24, no. 3 (2015): 256–71; and Nicole Wetsman, "Hospitals Are Selling Treasure Troves of Medical Data—What Could Go Wrong," The Verge, June 23, 2021, https://www.theverge.com/2021/6/23/22547397/medical-records-health-data-hospitals-research.

32. Eric J. Cassell, "The Relief of Suffering" *Archives of Internal Medicine* 143, no. 3 (1983): 522–23; and Timothy E. Quill and Christine K. Cassel, "Nonabandonment: A Central Obligation for Physicians," *Annals of Internal Medicine* 122, no. 5 (1995): 368–74.

33. Charles E. Binkley, "How Should Surgeons Communicate about Palliative and Curative Intentions, Purposes, and Outcomes?," *AMA Journal of Ethics* 23, no. 10 (2021): 794–99.

34. Raja Parasuraman and Dietrich H. Manzey, "Complacency and Bias in Human Use of Automation: An Attentional Integration," *Human Factors* 52, no. 3 (2010): 381–410.

35. Daisuke Wakabayashi, "Google and the University of Chicago Are Sued over Data Sharing," *New York Times*, June 26, 2019.

36. Melissa McCradden, James A. Anderson, Elizabeth A. Stephenson, Erik Drysdale, Lauren Erdman, Anna Goldenberg, and Randi Zlotnik Shaul, "A Research Ethics Framework for the Clinical Translation of Healthcare Machine Learning," *American Journal of Bioethics* 22, no. 5 (2022): 8–22.

37. Ma'N. Zawati and Michael Lang, "What's in the Box? Uncertain Accountability of Machine Learning Applications in Healthcare," *American Journal of Bioethics* 20, no. 11 (2020): 37–40.

38. W. Nicholson Price, Sara Gerke, and I. Glenn Cohen, "Potential Liability for Physicians Using Artificial Intelligence," *JAMA* 322, no. 18 (2019): 1765–66.

39. Brian K. Cooke, Elizabeth Worsham, and Gary M. Reisfield, "The Elusive Standard of Care," *Journal of the American Academy of Psychiatry and Law* 45, no. 3 (2017): 358–64.

40. Hannah Bleher and Matthias Braun. "Diffused Responsibility: Attributions of Responsibility in the Use of AI-Driven Clinical Decision Support Systems," *AI and Ethics* 2, no. 4 (2022): 747–61.

41. Bleher and Braun, "Diffused Responsibility," 747–61.

42. Min Kyung Lee, Daniel Kusbit, Anson Kahng, Ji Tae Kim, Xinran Yuan, Allissa Chan, Daniel See, et al. "WeBuildAI: Participatory Framework for Algorithmic Governance," *Proceedings of the ACM on Human-Computer Interaction* 3, no. CSCW (2019): 1–35.

4. THE DEVELOPER AND AI CLINICAL DECISION SUPPORT SYSTEMS

1. Amir Rosenfeld, Richard Zemel, and John K. Tsotsos, "The Elephant in the Room," arXiv, August 9, 2018, https://doi.org/10.48550.arXiv.1808.03305.

2. Scott M. Lundberg and Su-In Lee, "A Unified Approach to Interpreting Model Predictions," *Advances in Neural Information Processing Systems* 30 (2017): 4768–77.

3. Chad G. Ball et al., "The Impact of Shorter Prehospital Transport Times on Outcomes in Patients with Abdominal Vascular Injuries," *Journal of Trauma and Management Outcomes* 7, no. 1 (2013): 1–4.

4. Xiaojun Ma, Takeshi Imai, Emiko Shinohara, Satoshi Kasai, Kosuke Kato, Rina Kagawa, and Kazuhiko Ohe, "Ehr2ccas: A Framework for Mapping Ehr to Disease Knowledge Presenting Causal Chain of Disorders—Chronic Kidney Disease Example," *Journal of Biomedical Informatics* 115 (2021): 103692; Samantha Kleinberg and George Hripcsak, "A Review of Causal Inference for Biomedical Informatics," *Journal of Biomedical Informatics* 44, no. 6 (2011): 1102–12; Mattia Prosperi, Yi Guo, and Jiang Bian, "Bagged Random Causal Networks for Interventional Queries on Observational Biomedical Datasets," *Journal of Biomedical Informatics* 115 (2021): 103689; and Daniel C. Castro, Ian Walker, and Ben Glocker, "Causality Matters in Medical Imaging," *Nature Communications* 11, no. 1 (2020): 3673.

5. Yarin Gal and Zoubin Ghahramani, "Dropout as a Bayesian Approximation: Representing Model Uncertainty in Deep Learning."

Paper presented at the International Conference on Machine Learning, New York, June 19–24, 2016.

6. Kimberly E. Kopecky, David Urbach, and Margaret L. Schwarze, "Risk Calculators and Decision Aids Are Not Enough for Shared Decision Making," *JAMA Surgery* 154, no. 1 (2019): 3–4.; France Légaré, Stéphane Ratté, Karine Gravel, and Ian D. Graham, "Barriers and Facilitators to Implementing Shared Decision-Making in Clinical Practice: Update of a Systematic Review of Health Professionals' Perceptions," *Patient Education and Counseling* 73, no. 3 (2008): 526–35; and Tyler J. Loftus, Amanda C. Filiberto, Yanjun Li, Jeremy Balch, Allyson C. Cook, Patrick J. Tighe, Philip A. Efron, et al., "Decision Analysis and Reinforcement Learning in Surgical Decision-Making," *Surgery* 168, no. 2 (2020): 253–66.

7. Paul Grootendorst, David Feeny, and William Furlong, "Health Utilities Index Mark 3: Evidence of Construct Validity for Stroke and Arthritis in a Population Health Survey," *Medical Care* 38, no. 3 (2000): 290–99; Paul P. Glasziou, Sharon Bromwich, and R. John Simes, "Quality of Life Six Months after Myocardial Infarction Treated with Thrombolytic Therapy," *Medical Journal of Australia* 161, no. 9 (1994): 532–36; Neil A. Solomon, Henry A. Glick, Christopher J. Russo, Jason Lee, and Kevin A. Schulman, "Patient Preferences for Stroke Outcomes," *Stroke* 25, no. 9 (1994): 1721 25; and Malcolm Man-Son-Hing, Andreas Laupacis, Annette O'Connor, George Wells, Jacques Lemelin, William Wood, and Mark Dermer, "Warfarin for Atrial Fibrillation: The Patient's Perspective," *Archives of Internal Medicine* 156, no. 16 (1996): 1841–48.

8. Crystel M. Gijsberts, Karlijn A. Groenewegen, Imo E. Hoefer, Marinus J. C. Eijkemans, Folkert W. Asselbergs, Todd J. Anderson, Annie R. Britton, et al., "Race/Ethnic Differences in the Associations of the Framingham Risk Factors with Carotid IMT and Cardiovascular Events," *PLoS One* 10, no. 7 (2015): e0132321.

9. Adewole S. Adamson and Avery Smith, "Machine Learning and Health Care Disparities in Dermatolog,." *JAMA Dermatology* 154, no. 11 (2018): 1247–48; Ruha Benjamin, "Assessing Risk, Automating Racism," *Science* 366, no. 6464 (2019): 421–22; Ziad Obermeyer,

Brian Powers, Christine Vogeli, and Sendhil Mullainathan, "Dissecting Racial Bias in an Algorithm Used to Manage the Health of Populations," *Science* 366, no. 6464 (2019): 447–53; Caroline Criado Perez, *Invisible Women: Data Bias in a World Designed for Men* (New York: Abrams, 2019); and Alice B. Popejoy, Deborah I. Ritter, Kristy Crooks, Erin Currey, Stephanie M. Fullerton, Lucia A. Hindorff, Barbara Koenig, et al., "The Clinical Imperative for Inclusivity: Race, Ethnicity, and Ancestry (Rea) in Genomics," *Human Mutation* 39, no. 11 (2018): 1713–20.

10. Charles E. Binkley and David S. Kemp, "Ethical Centralization of High-Risk Surgery Requires Racial and Economic Justice," *Annals of Surgery* 272, no. 6 (2020): 917–18.

11. David M. Shahian, Jeffrey P. Jacobs, Vinay Badhwar, Paul A. Kurlansky, Anthony P. Furnary, Joseph C. Cleveland Jr., Kevin W. Lobdell, et al., "The Society of Thoracic Surgeons 2018 Adult Cardiac Surgery Risk Models: Part 1—Background, Design Considerations, and Model Development," *Annals of Thoracic Surgery* 105, no. 5 (2018): 1411–18; and Darshali A. Vyas, Leo G. Eisenstein, and David S. Jones, "Hidden in Plain Sight—Reconsidering the Use of Race Correction in Clinical Algorithms," *New England Journal of Medicine* 383, no. 9 (2020): 874–82.

12. Tezcan Ozrazgat-Baslanti et al., "Development and Validation of a Race-Agnostic Computable Phenotype for Kidney Health in Adult Hospitalized Patients," *Frontiers in Artificial Intelligence*, forthcoming.

13. Fatemeh Amrollahi, Supreeth P. Shashikumar, Angela Meier, Lucila Ohno-Machado, Shamim Nemati, and Gabriel Wardi, "Inclusion of Social Determinants of Health Improves Sepsis Readmission Prediction Models," *Journal of the American Medical Informatics Association* 29, no. 7 (2022): 1263–70.

14. Tom L. Beauchamp and James F. Childress, *Principles of Biomedical Ethics*, 8th ed. (New York: Oxford University Press, 2019).

15. Lukas J. Meier, Alice Hein, Klaus Diepold, and Alena Buyx, "Algorithms for Ethical Decision-Making in the Clinic: A Proof of Concept," *American Journal of Bioethics* 22, no. 7 (2022): 4–20.

16. Meier et al., "Algorithms for Ethical Decision-Making in the Clinic."

5. THE HEALTH SYSTEM EXECUTIVE AND AI CLINICAL DECISION SUPPORT SYSTEMS

1. Zoë Slote Morris, Steven Wooding, and Jonathan Grant, "The Answer Is 17 Years, What Is the Question: Understanding Time Lags in Translational Research," *Journal of the Royal Society of Medicine* 104, no. 12 (2011): 510–20.; Stephen T. Holgate, "The Future of Lung Research in the UK," *Thorax* 62, no 12 (2007): 1028–32; E. Andrew Balas and Suzanne A. Boren, "Managing Clinical Knowledge for Health Care Improvement," *Yearbook of Medical Informatics* 9, no. 1 (2000): 65–70; Health Economics Research Group, Office of Health Economics, and RAND Europe, *Medical Research: What's It Worth?*, November 2008, www.ukri.org/wp-content/uploads/2022/02/MRC-030222-medical-research-whats-it-worth.pdf; and Katharina Wratschko, *Strategic Orientation and Alliance Portfolio Configuration* (Wiesbaden: Springer, 2009).

2. Jerry Avorn and Aaron S. Kesselheim, "Up Is Down—Pharmaceutical Industry Caution Vs. Federal Acceleration of COVID-19 Vaccine Approval," *New England Journal of Medicine* 383, no. 18 (2020): 1706–8.

3. U.S. Food and Drug Administration, "Guidance for Industry: E9 Statistical Principles for Clinical Trials," accessed November 22, 2022, www.fda.gov/regulatory-information/search-fda-guidance-documents/e9-statistical-principles-clinical-trials.

4. Leah Isakov, Andrew W. Lo, and Vahid Montazerhodjat, "Is the FDA Too Conservative or Too Aggressive? A Bayesian Decision Analysis of Clinical Trial Design," *Journal of Econometrics* 211, no. 1 (2019): 117–36.

5. "NCA—Intracranial Stenting and Angioplasty (Cag-00085r2)—Decision Memo," accessed November 22, 2022, www.cms.gov/medicare-coverage-database/view/ncacal-decision-memo.aspx?proposed=Y&ncaid=177.

6. Colin P. Derdeyn and Marc I. Chimowitz, "Angioplasty and Stenting for Atherosclerotic Intracranial Stenosis: Rationale for a

Randomized Clinical Trial," *Neuroimaging Clinics of North America* 17, no. 3 (2007): 355–63.

7. Marc I. Chimowitz, Michael J. Lynn, Colin P. Derdeyn, Tanya N. Turan, David Fiorella, Bethany F. Lane, L. Scott, et al., "Stenting versus Aggressive Medical Therapy for Intracranial Arterial Stenosis," *New England Journal of Medicine* 365, no. 11 (2011): 993–1003.

8. David W. Shimabukuro, Christopher W. Barton, Mitchell D. Feldman, Samson J. Mataraso, and Ritankar Das, "Effect of a Machine Learning-Based Severe Sepsis Prediction Algorithm on Patient Survival and Hospital Length of Stay: A Randomised Clinical Trial," *BMJ Open Respiratory Research* 4, no. 1 (2017): e000234.

9. Marije Wijnberge, Bart F. Geerts, Liselotte Hol, Nikki Lemmers, Marijn P. Mulder, Patrick Berge, Jimmy Schenk, et al., "Effect of a Machine Learning–Derived Early Warning System for Intraoperative Hypotension vs. Standard Care on Depth and Duration of Intraoperative Hypotension During Elective Noncardiac Surgery: The Hype Randomized Clinical Trial," *JAMA* 323, no. 11 (2020): 1052–60, https://doi.org/10.1001/jama.2020.0592; and Ward H. van der Ven, Denise P. Veelo, Marije Wijnberge, Björn J. P. van der Ster, Alexander P. J. Vlaar, and Bart F. Geerts, "One of the First Validations of an Artificial Intelligence Algorithm for Clinical Use: The Impact on Intraoperative Hypotension Prediction and Clinical Decision-Making," *Surgery* 169, no. 6 (June 1, 2021): 1300–1303, https://doi.org/https://doi.org/10.1016/j.surg.2020.09.041, https://www.sciencedirect.com/science/article/pii/S0039606020307728.

10. J. M. Loree, S. Anand, A. Dasari, J. M. Unger, A. Gothwal, L. M. Ellis, et al., "Disparity of Race Reporting and Representation in Clinical Trials Leading to Cancer Drug Approvals from 2008 to 2018," *JAMA Oncology* 5, no. 10 (2019): e191870; and T. J. Moore, H. Zhang, G. Anderson, and G. C. Alexander, "Estimated Costs of Pivotal Trials for Novel Therapeutic Agents Approved by the US Food and Drug Administration, 2015–2016," *JAMA Internal Medicine* 178, no. 11 (2018): 1451–57.

11. Ian Ford and John Norrie, "Pragmatic Trials," *New England Journal of Medicine* 375, no. 5 (2016): 454–63, https://doi.org/10.1056/NEJMra1510059; and Steven R. Steinhubl, Dana L. Wolff-Hughes, Wendy Nilsen, Erin Iturriaga, and Robert M Califf, "Digital Clinical

Trials: Creating a Vision for the Future." *NPJ Digital Medicine* 2, no. 1 (2019): 126.

12. Steinhubl et al., "Digital Clinical Trials," 126.

13. Jay A. Pandit, Jennifer M. Radin, Giorgio Quer, and Eric J. Topol, "Smartphone Apps in the COVID-19 Pandemic," *Nature Biotechnology* 40, no. 7 (2022): 1013–22; Tyler J. Loftus, Patrick J. Tighe, Amanda C. Filiberto, Philip A. Efron, Scott C. Brakenridge, Alicia M. Mohr, Parisa Rashidi, et al., "Artificial Intelligence and Surgical Decision-Making," *JAMA Surgery* 155, no. 2 (2020): 148–58; and Le Peng, Gaoxiang Luo, Andrew Walker, Zachary Zaiman, Emma K. Jones, Hemant Gupta, Kristopher Kersten, et al., "Evaluation of Federated Learning Variations for COVID-19 Diagnosis Using Chest Radiographs from 42 US and European Hospitals," *Journal of the American Medical Informatics Association* 30, no. 1 (2023): 54–63.

14. Lucas M. Fleuren, Patrick Thoral, Duncan Shillan, Ari Ercole, Paul W. G. Elbers, and Right Data Right Now Collaborators, "Machine Learning in Intensive Care Medicine: Ready for Take-Off?," *Intensive Care Medicine* 46 (2020): 1486–88.

15. Loftus et al., "Artificial Intelligence and Surgical Decision-Making," 148–58.

16. Joaquin Quinonero-Candela, Masashi Sugiyama, Anton Schwaighofer, and Neil D. Lawrence, *Dataset Shift in Machine Learning* (Cambridge, MA: MIT Press, 2008); and Adarsh Subbaswamy and Suchi Saria, "From Development to Deployment: Dataset Shift, Causality, and Shift-Stable Models in Health AI," *Biostatistics* 21, no. 2 (2020): 345–52.

17. Samuel G. Finlayson, Adarsh Subbaswamy, Karandeep Singh, John Bowers, Annabel Kupke, Jonathan Zittrain, Isaac S. Kohane, and Suchi Saria, "The Clinician and Dataset Shift in Artificial Intelligence," *New England Journal of Medicine* 385, no. 3 (2021): 283–86.

18. Ittai Dayan, Holger R. Roth, Aoxiao Zhong, Ahmed Harouni, Amilcare Gentili, Anas Z. Abidin, Andrew Liu, et al., "Federated Learning for Predicting Clinical Outcomes in Patients with COVID-19," *Nature Medicine* 27, no. 10 (2021): 1735–43; and Peng et al., "Evaluation of Federated Learning Variations," 54–63.

19. Luca Melis, Congzheng Song, Emiliano De Cristofaro, and Vitaly Shmatikov, "Exploiting Unintended Feature Leakage in Collaborative Learning" (paper presented at the 2019 IEEE Symposium on Security and Privacy [SP], San Franciso, May 23, 2019); Milad Nasr, Reza Shokri, and Amir Houmansadr, "Comprehensive Privacy Analysis of Deep Learning: Passive and Active White-Box Inference Attacks against Centralized and Federated Learning" (paper presented at the 2019 IEEE Symposium on Security and Privacy [SP], San Franciso, May 23, 2019) Wenqi Wei, Ling Liu, Margaret Loper, Ka-Ho Chow, Mehmet Emre Gursoy, Stacey Truex, and Yanzhao Wu, "A Framework for Evaluating Gradient Leakage Attacks in Federated Learning," arXiv, April 23, 2020, https://doi.org/10.48550/arXiv.2004.10397; Briland Hitaj, Giuseppe Ateniese, and Fernando Perez-Cruz, "Deep Models under the Gan: Information Leakage from Collaborative Deep Learning" (paper presented at the Proceedings of the 2017 ACM SIGSAC Conference on Computer and Communications Security, Dallas, October 30–November 3, 2017); and Zhibo Wang, Mengkai Song, Zhifei Zhang, Yang Song, Qian Wang, and Hairong Qi, "Beyond Inferring Class Representatives: User-Level Privacy Leakage from Federated Learning" (paper presented at the IEEE INFOCOM 2019–IEEE Conference on Computer Communications, Paris, April 29–May 2, 2019.).

20. J. Andrew Onesimu, J. Karthikeyan, and Yuichi Sei, "An Efficient Clustering-Based Anonymization Scheme for Privacy-Preserving Data Collection in Iot Based Healthcare Services," *Peer-to-Peer Networking and Applications* 14 (2021): 1629–49; and Jiachun Li, Yan Meng, Lichuan Ma, Suguo Du, Haojin Zhu, Qingqi Pei, and Xuemin Shen, "A Federated Learning Based Privacy-Preserving Smart Healthcare System," *IEEE Transactions on Industrial Informatics* 18, no. 3 (2021), https://ieeexplore.ieee.org/stamp/stamp.jsp?arnumber=9492000.

6. INCORPORATING ETHICS INTO THE AI CLINICAL DECISION SUPPORT SYSTEM LIFE CYCLE

1. Matthias Braun, Patrik Hummel, Susanne Beck, and Peter Dabrock, "Primer on an Ethics of AI-Based Decision Support Systems

in the Clinic," *Journal of Medical Ethics* 47, no. 12 (2021): e3; Danton S. Char, Michael D. Abràmoff, and Chris Feudtner, "Identifying Ethical Considerations for Machine Learning Healthcare Applications," *American Journal of Bioethics* 20, no. 11 (2020): 7–17; Irene Y. Chen, Emma Pierson, Sherri Rose, Shalmali Joshi, Kadija Ferryman, and Marzyeh Ghassemi, "Ethical Machine Learning in Healthcare," *Annual Review of Biomedical Data Science* 4 (2021): 123–44; Luciano Floridi, Josh Cowls, Monica Beltrametti, Raja Chatila, Patrice Chazerand, Virginia Dignum, Christoph Luetge, et al., "An Ethical Framework for a Good AI Society: Opportunities, Risks, Principles, and Recommendations," in *Ethics, Governance, and Policies in Artificial Intelligence*, ed. Luciano Floridi (Cham, Switzerland: Springer Nature, 2021) 19–39; Melissa D. McCradden, James A. Anderson, Elizabeth A. Stephenson, Erik Drysdale, Lauren Erdman, Anna Goldenberg, and Randi Zlotnik Shaul, "A Research Ethics Framework for the Clinical Translation of Healthcare Machine Learning," *American Journal of Bioethics* 22, no. 5 (2022): 8–22; Jessica Morley, Caio C. V. Machado, Christopher Burr, Josh Cowls, Indra Joshi, Mariarosaria Taddeo, and Luciano Floridi, "The Ethics of AI in Health Care: A Mapping Review," *Social Science & Medicine* 260 (2020): 113–72; Sandeep Reddy, Wendy Rogers, Ville-Petteri Makinen, Enrico Coiera, Pieta Brown, Markus Wenzel, Eva Weicken, et al., "Evaluation Framework to Guide Implementation of AI Systems into Healthcare Settings," *BMJ Health & Care Informatics* 28, no. 1 (2021) e100444; Sebastian Vollmer, Bilal A. Mateen, Gergo Bohner, Franz J. Király, Rayid Ghani, Pall Jonsson, Sarah Cumbers, et al., "Machine Learning and Artificial Intelligence Research for Patient Benefit: 20 Critical Questions on Transparency, Replicability, Ethics, and Effectiveness," *BMJ* 368 (2020): l6927; Jenna Wiens, Suchi Saria, Mark Sendak, Marzyeh Ghassemi, Vincent X. Liu, Finale Doshi-Velez, Kenneth Jung, et al., "Do No Harm: A Roadmap for Responsible Machine Learning for Health Care," *Nature Medicine* 25, no. 9 (2019): 1337–40; and Roberto V. Zicari, John Brodersen, James Brusseau, Boris Düdder, Timo Eichhorn, Todor Ivanov, Georgios Kararigas, et al., "Z-Inspection®: A Process to Assess Trustworthy AI," *IEEE Transactions on Technology and Society* 2, no. 2 (2021): 83–97.

2. Charles E. Binkley, Michael S. Politz, and Brian P. Green, "Who, If Not the FDA, Should Regulate Implantable Brain-Computer Interface Devices?," *AMA Journal of Ethics* 23, no. 9 (2021): 745–49.

3. Virgílio Almeida, Laura Schertel Mendes, and Danilo Doneda, "On the Development of AI Governance Frameworks," *IEEE Internet Computing* 27, no. 1 (2023): 70–74.

4. Joseph Donia and James A. Shaw, "Co-design and Ethical Artificial Intelligence for Health: An Agenda for Critical Research and Practice," *Big Data & Society* 8, no. 2 (2021): 20539517211065248; and Min Kyung Lee, Daniel Kusbit, Anson Kahng, Ji Tae Kim, Xinran Yuan, Allissa Chan, Daniel See, et al., "WeBuildAI: Participatory Framework for Algorithmic Governance," *Proceedings of the ACM on Human-Computer Interaction* 3, item CSCW (2019): 1–35.

5. Reid Blackman, "Why You Need an AI Ethics Committee: Expert Oversight Will Help You Safe-Guard Your Data and Your Brand," *Harvard Business Review* 100, no. 7–8 (2022): 118–25.

6. Phoebe Friesen, Rachel Douglas-Jones, Mason Marks, Robin Pierce, Katherine Fletcher, Abhishek Mishra, Jessica Lorimer, et al., "Governing AI-Driven Health Research: Are IRBs Up to the Task?," *Ethics & Human Research* 43, no. 2 (2021): 35–42.

7. Stephanie Eaneff, Ziad Obermeyer, and Atul J. Butte, "The Case for Algorithmic Stewardship for Artificial Intelligence and Machine Learning Technologies," *JAMA* 324, no. 14 (2020): 1397–98.

8. Jee Young Kim, William Boag, Freya Gulamali, Alifia Hasan, Henry David Jeffry Hogg, Mark Lifson, Deirdre Mulligan, et al., "Organizational Governance of Emerging Technologies: AI Adoption in Healthcare," in *Proceedings of the 2023 ACM Conference on Fairness, Accountability, and Transparency* (New York, NY: Association for Computing Machinery, 2023), 1396–1417.

9. Melissa D. McCradden, James A. Anderson, Elizabeth A. Stephenson, Erik Drysdale, Lauren Erdman, Anna Goldenberg, and Randi Zlotnik Shaul, "A Research Ethics Framework for the Clinical Translation of Healthcare Machine Learning," *American Journal of Bioethics* 22, no. 5 (2022): 8–22.

10. Regina G. Russell, Laurie Lovett Novak, Mehool Patel, Kim V. Garvey, Kelly Jean Thomas Craig, Gretchen P. Jackson, Don Moore, et al., "Competencies for the Use of Artificial Intelligence–Based Tools by Health Care Professionals," *Academic Medicine* 98, no. 3 (2023): 348–56.

11. Kim et al., "Organizational Governance of Emerging Technologies," 1396–1417.

12. McCradden et al., "Research Ethics Framework," 8–22.

13. McCradden et al., "Research Ethics Framework," 8–22.

14. Kim et al., "Organizational Governance of Emerging Technologies," 1396–1417.

15. Char et al., "Identifying Ethical Considerations," 7–17.

16. Char et al., "Identifying Ethical Considerations," 7–17.

17. Char et al., "Identifying Ethical Considerations," 7–17.

18. Char et al., "Identifying Ethical Considerations," 7–17.

19. Char et al., "Identifying Ethical Considerations," 7–17.

20. Juan C. Rojas, John Fahrenbach, Sonya Makhni, Scott C. Cook, James S. Williams, Craig A. Umscheid, and Marshall H. Chin, "Framework for Integrating Equity into Machine Learning Models: A Case Study," *Chest* 161, no. 6 (2022): 1621–27.

21. Juan C. Rojas, Mario Teran, and Craig A. Umscheid, "Clinician Trust in Artificial Intelligence: What Is Known and How Trust Can Be Facilitated," *Critical Care Clinics* 39, no. 4 (2023): 769–82.

22. Kim et al., "Organizational Governance of Emerging Technologies," 1396–1417.

23. Char et al., "Identifying Ethical Considerations," 7–17.

24. Rojas et al., "Framework for Integrating Equity," 1621–27.

25. Char et al., "Identifying Ethical Considerations," 7–17; Kristin M. Kostick-Quenet, I. Glenn Cohen, Sara Gerke, Bernard Lo, James Antaki, Faezah Movahedi, Hasna Njah, et al. "Mitigating Racial Bias in Machine Learning," *Journal of Law, Medicine & Ethics* 50, no. 1 (2022): 92–100; and Wiens et al., "Do No Harm," 1337–40.

26. Char et al., "Identifying Ethical Considerations," 7–17; McCradden et al., "Research Ethics Framework," 8–22; and Wiens et al., "Do No Harm," 1337–40.

27. McCradden et al., "Research Ethics Framework," 8–22; and Wiens et al., "Do No Harm," 1337–40.

28. McCradden et al., "Research Ethics Framework," 8–22.

29. Russell et al., "Use of Artificial Intelligence–Based Tools," *Academic Medicine* 98, no. 3 (2023): 348–56; and Kim et al., "Organizational Governance of Emerging Technologies," 1396–1417.

30. Rojas et al., "Clinician Trust in Artificial Intelligence."

31. The Data Nutrition Project, home page, accessed September 25, 2023, https://datanutrition.org/.

32. Char et al., "Identifying Ethical Considerations," 7–17; McCradden et al., "Research Ethics Framework," 8–22; and Kim et al., "Organizational Governance of Emerging Technologies," 1396–1417.

33. Ravi B. Parikh, and Lorens A. Helmchen. "Paying for Artificial Intelligence in Medicine," *NPJ Digital Medicine* 5, no. 1 (2022): 63.

34. Charles Binkley, "The Physician's Conundrum: Assigning Moral Responsibility for Medical Artificial Intelligence and Machine Learning," *Verdict*, February 8, 2021, https://verdict.justia.com/2021/02/08/the-physicians-conundrum; and Charles E. Binkley and Brian P. Green, "Does Intraoperative Artificial Intelligence Decision Support Pose Ethical Issues?," *JAMA Surgery* 156, no. 9 (2021): 809–10.

35. Jennifer Blumenthal-Barby, "An AI Bill of Rights: Implications for Health Care AI and Machine Learning—A Bioethics Lens," *American Journal of Bioethics* 23, no. 1 (2023): 4–6.

36. Kim et al., "Organizational Governance of Emerging Technologies," 1396–1417.

37. Char et al., "Identifying Ethical Considerations," 7–17; and Kim et al., "Organizational Governance of Emerging Technologies," 1396–1417.

38. Char et al., "Identifying Ethical Considerations," 7–17; and Kim et al., "Organizational Governance of Emerging Technologies," 1396–1417.

39. Hannah Bleher and Matthias Braun, "Diffused Responsibility: Attributions of Responsibility in the Use of AI-Driven Clinical Decision Support Systems," *AI and Ethics* 2, no. 4 (2022): 747–61; and World Health Organization, "Ethics and Governance of Artificial

Intelligence for Health," *WHO Guidance*, June 28, 2021, www.who.int/publications/i/item/9789240029200.

40. A. D. Stern, A. Goldfarb, T. Minssen, and W. N. Price II, "AI Insurance: How Liability Insurance Can Drive the Responsible Adoption of Artificial Intelligence in Health Care," *NEJM Catalyst Innovations in Care Delivery* 3, no. 4 (2022): CAT-21.

BIBLIOGRAPHY

Adamson, Adewole S., and Avery Smith. "Machine Learning and Health Care Disparities in Dermatology." *JAMA Dermatology* 154, no. 11 (2018): 1247–48.

Almeida, Virgílio, Laura Schertel Mendes, and Danilo Doneda. "On the Development of AI Governance Frameworks." *IEEE Internet Computing* 27, no. 1 (2023): 70–74.

Amann, Julia, Alessandro Blasimme, Effy Vayena, Dietmar Frey, and Vince I. Madai. "Explainability for Artificial Intelligence in Healthcare: A Multidisciplinary Perspective." *BMC Medical Informatics and Decision Making* 20, no. 1 (2020): 1–9.

Amrollahi, Fatemeh, Supreeth P. Shashikumar, Angela Meier, Lucila Ohno-Machado, Shamim Nemati, and Gabriel Wardi. "Inclusion of Social Determinants of Health Improves Sepsis Readmission Prediction Models." *Journal of the American Medical Informatics Association* 29, no. 7 (2022): 1263–70.

Angus, Derek C. "Randomized Clinical Trials of Artificial Intelligence." *JAMA* 323, no. 11 (2020): 1043–45.

Avorn, Jerry, and Aaron S. Kesselheim. "Up Is Down—Pharmaceutical Industry Caution vs. Federal Acceleration of COVID-19 Vaccine Approval." *New England Journal of Medicine* 383, no. 18 (2020): 1706–8.

Balas, E. Andrew, and Suzanne A Boren. "Managing Clinical Knowledge for Health Care Improvement." *Yearbook of Medical Informatics* 9, no. 1 (2000): 65–70.

Ball, Chad G., Brian H. Williams, C. Tallah, Jeffrey P. Salomone, and David V. Feliciano. "The Impact of Shorter Prehospital Transport Times on Outcomes in Patients with Abdominal Vascular Injuries." *Journal of Trauma and Management Outcomes* 7, no. 1 (2013): 1–4.

Banja, John. "How Might Artificial Intelligence Applications Impact Risk Management?" *AMA Journal of Ethics* 22, no. 11 (2020): 945–51.

Barrow, Jennifer M., Grace D. Brannan, and Paras B. Khandhar. "Research Ethics." In *StatPearls* (Treasure Island, FL: StatPearls Publishing; 2023). https://www.ncbi.nlm.nih.gov/books/NBK459281/.

Basson, Marc D., Gerald Dworkin, and Eric J. Cassell. "The 'Student Doctor' and a Wary Patient." *Hastings Center Report* 12, no. 1 (1982): 27–28.

Beauchamp, Tom L., and James F. Childress.*Principles of Biomedical Ethics*. 8th ed. New York: Oxford University Press, 2019.

Becker, Gary J. "Human Subjects Investigation: Timeless Lessons of Nuremberg and Tuskegee." *Journal of the American College of Radiology* 2, no. 3 (2005): 215–17.

Benjamin, Ruha. "Assessing Risk, Automating Racism." *Science* 366, no. 6464 (2019): 421–22.

Beskow, Laura M. "Lessons from HeLa Cells: The Ethics and Policy of Biospecimens." *Annual Review of Genomics and Human Genetics* 17 (2016): 395–417.

Bilimoria, Karl Y., Yaoming Liu, Jennifer L. Paruch, Lynn Zhou, Thomas E. Kmiecik, Clifford Y. Ko, and Mark E. Cohen. "Development and Evaluation of the Universal ACS NSQIP Surgical Risk Calculator: A Decision Aid and Informed Consent Tool for Patients and Surgeons." *Journal of the American College of Surgeons* 217, no. 5 (2013): 833–42.

Binkley, Charles. "The Physician's Conundrum: Assigning Moral Responsibility for Medical Artificial Intelligence and Machine Learning." *Verdict*, February 8, 2021. https://verdict.justia.com/2021/02/08/the-physicians-conundrum.

Binkley, Charles E. "How Should Surgeons Communicate about Palliative and Curative Intentions, Purposes, and Outcomes?" *AMA Journal of Ethics* 23, no. 10 (2021): 794–99.

Binkley, Charles E., and Brian P. Green. "Does Intraoperative Artificial Intelligence Decision Support Pose Ethical Issues?" *JAMA Surgery* 156, no. 9 (2021): 809–10.

Binkley, Charles E., and David S. Kemp. "Ethical Centralization of High-Risk Surgery Requires Racial and Economic Justice." *Annals of Surgery* 272, no. 6 (2020): 917–18.

Binkley, Charles E., David S. Kemp, and Brandi Braud Scully. "Should We rely on AI to Help Avoid Bias in Patient Selection for Major Surgery?" *AMA Journal of Ethics* 24, no. 8 (2022): 773–80.

Binkley, Charles E., and Bryan Pilkington. "The Actionless Agent: An Account of Human-CAI Relationships." *American Journal of Bioethics* 23, no. 5 (2023): 25–27.

Binkley, Charles E., and Bryan Pilkington. "Informed Consent for Clinician-AI Collaboration and Patient Data Sharing: Substantive, Illusory, or Both?" *American Journal of Bioethics* 23, no. 10 (2023): 83–85.

Binkley, Charles E., Michael S. Politz, and Brian P. Green. "Who, If Not the FDA, Should Regulate Implantable Brain-Computer Interface Devices?" *AMA Journal of Ethics* 23, no. 9 (2021): 745–49.

Binkley, Charles E., Joel Michael Reynolds, and Andrew Shuman. "From the Eyeball Test to the Algorithm—Quality of Life, Disability Status, and Clinical Decision Making in Surgery." *New England Journal of Medicine* 387, no. 14 (2022): 1325–28.

Birhane, Abeba, Pratyusha Kalluri, Dallas Card, William Agnew, Ravit Dotan, and Michelle Bao. "The Values Encoded in Machine Learning Research." In *FaccT '22: Proceedings of the 2022 ACM Conference on Fairness, Accountability, and Transparency*, 173–84. Association for Computing Machinery, 2022. https://dl.acm.org/doi/10.1145/3531146.3533083.

Blackman, Reid "Why You Need an AI Ethics Committee: Expert Oversight Will Help You Safe-Guard Your Data and Your brand." *Harvard Business Review* 100, nos. 7–8 (2022): 118–25.

Bleher, Hannah, and Matthias Braun. "Diffused responsibility: attributions of responsibility in the use of AI-driven clinical decision support systems." *AI and Ethics* 2, no. 4 (2022): 747–761

Blumenthal-Barby, Jennifer "An AI Bill of Rights: Implications for Health Care AI and Machine Learning—A Bioethics Lens." *American Journal of Bioethics* 23, no. 1 (2023): 4–6.

Braun, Matthias, Patrik Hummel, Susanne Beck, and Peter Dabrock. "Primer on an Ethics of AI-Based Decision Support Systems in the Clinic." *Journal of Medical Ethics* 47, no. 12 (2021): e3.

Carter, Stacy M., Wendy Rogers, Khin Than Win, Helen Frazer, Bernadette Richards, and Nehmat Houssami. "The Ethical, Legal and Social Implications of Using Artificial Intelligence Systems in Breast Cancer Care." *Breast* 49 (2020): 25–32.

Cassell, Eric J. "The Relief of Suffering," *Archives of Internal* Medicine 143, no. 3 (1983): 522–23.

Castro, Daniel C., Ian Walker, and Ben Glocker. "Causality Matters in Medical Imaging." *Nature Communications* 11, no. 1 (2020): 3673.

Cavanaugh, Thomas A. *Hippocrates' Oath and Asclepius' Snake: The Birth of the Medical Profession.* New York: Oxford University Press, 2018.

Challen, Robert, Joshua Denny, Martin Pitt, Luke Gompels, Tom Edwards, and Krasimira Tsaneva-Atanasova. "Artificial Intelligence, Bias and Clinical Safety." *BMJ Quality & Safety* 28, no. 3 (2019): 231–37.

Char, Danton S., Michael D. Abràmoff, and Chris Feudtner. "Identifying Ethical Considerations for Machine Learning Healthcare Applications." *American Journal of Bioethics* 20, no. 11 (2020): 7–17.

Char, Danton S., Nigam H. Shah, and David Magnus. "Implementing Machine Learning in Health care—Addressing Ethical Challenges." *New England Journal of Medicine* 378, no. 11 (2018): 981.

Chen, Irene Y., Emma Pierson, Sherri Rose, Shalmali Joshi, Kadija Ferryman, and Marzyeh Ghassemi. "Ethical Machine Learning in Healthcare." *Annual Review of Biomedical Data Science* 4 (2021): 123–44.

Chen, Stacy S., and Sunit Das. "What Are My Options?": Physicians as Ontological Decision Architects in Surgical Informed Consent." *Bioethics* 36, no. 9 (2022): 936–39.

Chimowitz, Marc I., Michael J. Lynn, Colin P. Derdeyn, Tanya N. Turan, David Fiorella, Bethany F. Lane, L. Scott Janis, et al. "Stenting versus Aggressive Medical Therapy for Intracranial Arterial Stenosis." *New England Journal of Medicine* 365, no. 11 (2011): 993–1003.

Clouser, K. Danner, and Bernard Gert. "A Critique of Principlism." *Journal of Medicine and Philosophy* 15, no. 2 (1990): 219–36.

Cohen, Daniel L,. Laurence B. McCullough, R. W. Kessel, Aristide Y. Apostolides, Errol R. Alden, and Kelly J. Heiderich. "Informed Consent Policies Governing Medical Students' Interactions with Patients." *Academic Medicine* 62, no. 10 (1987): 789–98.

Cohen, I. Glenn. "Informed Consent and Medical Artificial Intelligence: What to Tell the Patient?" *Georgetown Law Journal* 108 (2019): 1425.

Coiera, Enrico "The Fate of Medicine in the Time of AI." *Lancet* 392, no. 10162 (2018): 2331–32.

Cooke, Brian K., Elizabeth Worsham, and Gary M. Reisfield. "The Elusive Standard of Care." *Journal of the American Academy of Psychiatry and the Law* 45, no. 3 (2017): 358–64.

Dana, Jason, and George Loewenstein. "A Social Science Perspective on Gifts to Physicians from Industry." *JAMA* 290, no. 2 (2003): 252–55.

The Data Nutrition Project. Home page. Accessed September 25, 2023. https://datanutrition.org/.

Dayan, Ittai, Holger R. Roth, Aoxiao Zhong, Ahmed Harouni, Amilcare Gentili, Anas Z. Abidin, Andrew Liu, et al. "Federated Learning for Predicting Clinical Outcomes in Patients with COVID-19." *Nature Medicine* 27, no. 10 (2021): 1735–43.

Derdeyn, Colin P., and Marc I. Chimowitz. "Angioplasty and Stenting for Atherosclerotic Intracranial Stenosis: Rationale for a Randomized Clinical Trial." *Neuroimaging Clinics of North America* 17, no. 3 (2007): 355–63.

Donia, Joseph, and James A. Shaw. "Co-design and Ethical Artificial Intelligence for Health: An Agenda for Critical Research and Practice." *Big Data & Society* 8, no. 2 (2021): 20539517211065248.

Dorney, Maureen S. "Moore v. The Regents of the University of California: Balancing the Need for Biotechnology Innovation against the Right of Informed Consent." *High Technology Law Journal* 5 (1989): 333.

Duncan, Ian, Tamim Ahmed, Henry Dove, and Terri L. Maxwell. "Medicare Cost at End of Life." *American Journal of Hospice and Palliative Medicine* 36, no. 8 (2019): 705–10.

Eagle, Kim A., Peter B. Berger, Hugh Calkins, Bernard R. Chaitman, Gordon A. Ewy, Kirsten E. Fleischmann, Lee A. Fleisher, et al. "ACC/AHA Guideline Update for Perioperative Cardiovascular Evaluation for Noncardiac Surgery—Executive Summary: A Report of the American College of Cardiology/American Heart Association Task Force on Practice Guidelines (Committee to Update the 1996 Guidelines on Perioperative Cardiovascular Evaluation for Noncardiac Surgery)." *Journal of the American College of Cardiology* 39, no. 3 (2002): 542–53.

Eaneff, Stephanie, Ziad Obermeyer, and Atul J. Butte. "The Case for Algorithmic Stewardship for Artificial Intelligence and Machine Learning Technologies." *JAMA* 324, no. 14 (2020): 1397–98.

Emanuel, Ezekiel J., and Linda L. Emanuel. "Four Models of the Physician-Patient Relationship." *JAMA* 267, no. 16 (1992): 2221–26.

Finlayson, Samuel G., Adarsh Subbaswamy, Karandeep Singh, John Bowers, Annabel Kupke, Jonathan Zittrain, Isaac S. Kohane, and Suchi Saria. "The Clinician and Dataset Shift in Artificial Intelligence." *New England Journal of Medicine* 385, no. 3 (2021): 283–86.

Fleuren, Lucas M., Patrick Thoral, Duncan Shillan, Ari Ercole, Paul W. G. Elbers, and Right Data Right Now Collaborators. "Machine Learning in Intensive Care Medicine: Ready for Take-Off?" *Intensive Care Medicine* 46 (2020): 1486–88.

Floridi, Luciano, and Josh Cowls. "A Unified Framework of Five Principles for AI in Society." In *Machine Learning and the City: Applications in Architecture and Urban Design*, edited by Silvio Carta, 535–45. Wiley Online Library, 2022. https://doi.org/10.1002/9781119815075.ch45.

Floridi, Luciano, Josh Cowls, Monica Beltrametti, Raja Chatila, Patrice Chazerand, Virginia Dignum, Christoph Luetge, et al. "An Ethical Framework for a Good AI Society: Opportunities, Risks, Principles, and Recommendations." In *Ethics, Governance, and Policies in Artificial Intelligence*, edited by Luciano Floridi, 19–39. Cham, Switzerland: Springer Nature, 2021.

Ford, Ian, and John Norrie. "Pragmatic Trials." *New England Journal of Medicine* 375, no. 5 (2016): 454–63. https://doi.org/10.1056/NEJMra1510059. https://www.nejm.org/doi/full/10.1056/NEJMra1510059.

Fourie, Carina. "Sufficiency of Capabilities, Social Equality, and Two-Tiered Health Care Systems." In *What Is Enough? Sufficiency, Justice and Health*, edited by Carina Fourie and Anette Rid, 185–204. New York: Oxford University Press, 2017.

Friesen, Phoebe, Rachel Douglas-Jones, Mason Marks, Robin Pierce, Katherine Fletcher, Abhishek Mishra, Jessica Lorimer, et al. "Governing AI-Driven Health Research: Are IRBs Up to the Task?" *Ethics & Human Research* 43, no. 2 (2021): 35–42.

Gal, Yarin, and Zoubin Ghahramani. "Dropout as a Bayesian Approximation: Representing Model Uncertainty in Deep Learning." Paper presented at the International Conference on Machine Learning, New York, June 19–24, 2016.

Gijsberts, Crystel M., Karlijn A. Groenewegen, Imo E. Hoefer, Marinus J. C. Eijkemans, Folkert W. Asselbergs, Todd J. Anderson, Annie R. Britton, et al. "Race/Ethnic Differences in the Associations of the Framingham Risk Factors with Carotid IMT and Cardiovascular Events." *PLoS One* 10, no. 7 (2015): e0132321.

Glasziou, Paul P., Sharon Bromwich, and R. John Simes. "Quality of Life Six Months after Myocardial Infarction Treated with Thrombolytic Therapy." *Medical Journal of Australia* 161, no. 9 (1994): 532–36.

Gretton, Cosimi. "The Dangers of AI in Health Care: Risk Homeostasis and Automation Bias." *Towards Data Science* (blog), June 24, 2017. https://towardsdatascience.com/the-dangers-of-ai-in-health-care-risk-homeostasis-and-automation-bias-148477a9080f.

Grootendorst, Paul, David Feeny, and William Furlong. "Health Utilities Index Mark 3: Evidence of Construct Validity for Stroke and Arthritis in a Population Health Survey." *Medical Care* 38, no. 3 (2000): 290–99.

Grote, Thomas, and Philipp Berens. "On the Ethics of Algorithmic Decision-Making in Healthcare." *Journal of Medical Ethics*, 46, no. 3 (2020): 205–11.

Hacker, Philipp, Ralf Krestel, Stefan Grundmann, and Felix Naumann. "Explainable AI under Contract and Tort Law: Legal Incentives and technical Challenges." *Artificial Intelligence and Law* 28 (2020): 415–39.

Hallowell, Nina, Shirlene Badger, Aurelia Sauerbrei, Christoffer Nellåker, and Angeliki Kerasidou. "'I Don't Think People Are Ready to Trust These Algorithms at Face Value': Trust and the Use of Machine Learning algorithms in the Diagnosis of Rare Disease." *BMC Medical Ethics* 23, no. 1 (2022): 1–14.

Hammoud, Maya M., Kayte Spector-Bagdady, Meg O'Reilly, Carol Major, and Laura Baecher-Lind. "Consent for the Pelvic Examination under Anesthesia by Medical Students: Recommendations by the Association of Professors of Gynecology and Obstetrics." *Obstetrics and Gynecology* 134, no. 6 (2019): 1303.

Health Economics Research Group, Office of Health Economics, and RAND Europe. *Medical Research: What's It Worth?*, November 2008. www.ukri.org/wp-content/uploads/2022/02/MRC-030222-medical-research-whats-it-worth.pdf.

Hippocrates. *Hippocratic Writings.* Translated by G. E. R. Lloyd, John Chadwick, and W. N. Mann. New York: Penguin, 1978.

Hitaj, Briland, Giuseppe Ateniese, and Fernando Perez-Cruz. "Deep Models under the Gan: Information Leakage from Collaborative Deep Learning." Paper presented at the Proceedings of the 2017 ACM SIGSAC Conference on Computer and Communications Security, Dallas, October 30–November 3, 2017.

Holgate, Stephen T. "The Future of Lung Research in the UK." *Thorax* 62, no 12 (2007): 1028–32.

Isakov, Leah, Andrew W. Lo, and Vahid Montazerhodjat. "Is the FDA Too Conservative or Too Aggressive? A Bayesian Decision Analysis of Clinical Trial Design." *Journal of Econometrics* 211, no. 1 (2019): 117–36.

Jacobs, Maia, Melanie F. Pradier, Thomas H. McCoy Jr., Roy H. Perlis, Finale Doshi-Velez, and Krzysztof Z. Gajos. "How Machine-

Learning Recommendations Influence Clinician Treatment Selections: The Example of Antidepressant Selection." *Translational Psychiatry* 11, no. 1 (2021): 108.

Jagsi, Reshma, and Lisa Soleymani Lehmann. "The Ethics of Medical Education." *BMJ* 329, no. 7461 (2004): 332–34.

Johnson-Mann, Crystal N., Tyler J. Loftus, and Azra Bihorac. "Equity and Artificial Intelligence in Surgical Care." *JAMA Surgery* 156, no. 6 (2021): 509–10.

Kaplan, Bonnie. "Selling Health Data: De-identification, Privacy, and Speech." *Cambridge Quarterly of Healthcare Ethics* 24, no. 3 (2015): 256–71.

Kessler, Sharon E. "Why Care: Complex Evolutionary History of Human Healthcare Networks." *Frontiers in Psychology* 11 (2020): 199.

Kim, Jee Young, William Boag, Freya Gulamali, Alifia Hasan, Henry David Jeffry Hogg, Mark Lifson, Deirdre Mulligan, et al. "Organizational Governance of Emerging Technologies: AI Adoption in Healthcare." In *Proceedings of the 2023 ACM Conference on Fairness, Accountability, and Transparency*, 1396–1417. New York: Association for Computing Machinery, 2023.

Kleinberg, Samantha, and George Hripcsak. "A Review of Causal Inference for Biomedical Informatics." *Journal of Biomedical Informatics* 44, no. 6 (2011): 1102–12.

Kopecky, Kimberly E., David Urbach, and Margaret L Schwarze. "Risk Calculators and Decision Aids Are Not Enough for Shared Decision Making." *JAMA Surgery* 154, no. 1 (2019): 3–4.

Kostick-Quenet, Kristin M., I. Glenn Cohen, Sara Gerke, Bernard Lo, James Antaki, Faezah Movahedi, Hasna Njah, et al. "Mitigating Racial Bias in Machine Learning." *Journal of Law, Medicine & Ethics* 50, no. 1 (2022): 92–100.

Kraft, Stephanie A. "Respect and Trustworthiness in the Patient-Provider-Machine Relationship: Applying a Relational Lens to Machine Learning Healthcare Applications." *American Journal of Bioethics* 20, no. 11 (2020): 51–53.

Krugman, Saul. "The Willowbrook Hepatitis Studies Revisited: Ethical Aspects." *Reviews of Infectious Diseases* 8, no. 1 (1986): 157–62.

Lakkaraju, Himabindu, and Osbert Bastani. "'How Do I Fool You?' Manipulating User Trust via Misleading Black Box Explanations." In *AIES '20: Proceedings of the AAAI/ACM Conference on AI, Ethics, and Society* (February 2020). https://dl.acm.org/doi/10.1145/3375627.3375833.

Lee, Min Kyung, Daniel Kusbit, Anson Kahng, Ji Tae Kim, Xinran Yuan, Allissa Chan, Daniel See, et al. "WeBuildAI: Participatory Framework for Algorithmic Governance." *Proceedings of the ACM on Human-Computer Interaction* 3, item CSCW (2019): 1–35.

Légaré, France, Stéphane Ratté, Karine Gravel, and Ian D. Graham. "Barriers and Facilitators to Implementing Shared Decision-Making in Clinical Practice: Update of a Systematic Review of Health Professionals' Perceptions." *Patient Education and Counseling* 73, no. 3 (2008): 526–35.

Li, Jiachun, Yan Meng, Lichuan Ma, Suguo Du, Haojin Zhu, Qingqi Pei, and Xuemin Shen. "A Federated Learning Based Privacy-Preserving Smart Healthcare System." *IEEE Transactions on Industrial Informatics* 18, no. 3 (2021) https://ieeexplore.ieee.org/stamp/stamp.jsp?arnumber=9492000.

Liverman, Catharyn T., and James F. Childress, eds. *Organ Donation: Opportunities for Action.* Washington, DC: National Academies Press, 2006.

Lo, Bernard. *Resolving Ethical Dilemmas: A Guide for Clinicians.* 6th ed. Philadelphia, PA: Wolters Kluwer, 2020.

Loftus, Tyler J., Amanda C. Filiberto, Yanjun Li, Jeremy Balch, Allyson C. Cook, Patrick J. Tighe, Philip A. Efron, et al. "Decision Analysis and Reinforcement Learning in Surgical Decision-Making." *Surgery* 168, no. 2 (2020): 253–66.

Loftus, Tyler J., Patrick J. Tighe, Amanda C. Filiberto, Philip A. Efron, Scott C. Brakenridge, Alicia M. Mohr, Parisa Rashidi, et al. "Artificial Intelligence and Surgical Decision-Making." *JAMA Surgery* 155, no. 2 (2020): 148–58.

London, Alex John. "Artificial Intelligence and Black-Box Medical Decisions: Accuracy versus Explainability." *Hastings Center Report* 49, no. 1 (2019): 15–21.

Loree, J. M., S. Anand, A. Dasari, J. M. Unger, A. Gothwal, L. M. Ellis, et al. "Disparity of Race Reporting and Representation in Clinical Trials Leading to Cancer Drug Approvals from 2008 to 2018." *JAMA Oncology* 5, no. 10 (2019): e191870.

Lundberg, Scott M., and Su-In Lee. "A Unified Approach to Interpreting Model Predictions." *Advances in Neural Information Processing Systems* 30 (2017): 4768–77.

Ma, Xiaojun, Takeshi Imai, Emiko Shinohara, Satoshi Kasai, Kosuke Kato, Rina Kagawa, and Kazuhiko Ohe. "Ehr2ccas: A Framework for Mapping Ehr to Disease Knowledge Presenting Causal Chain of Disorders—Chronic Kidney Disease Example." *Journal of Biomedical Informatics* 115 (2021): 103692.

Man-Son-Hing, Malcolm, Andreas Laupacis, Annette O'Connor, George Wells, Jacques Lemelin, William Wood, and Mark Dermer. "Warfarin for Atrial Fibrillation: The Patient's Perspective." *Archives of Internal Medicine* 156, no. 16 (1996): 1841–48.

Marracino, Richelle K., and Robert D. Orr. "Entitling the Student Doctor: Defining the Student's Role in Patient Care." *Journal of General Internal Medicine* 13 (1998): 266–70.

McCradden, Melissa, James A. Anderson, Elizabeth A. Stephenson, Erik Drysdale, Lauren Erdman, Anna Goldenberg, and Randi Zlotnik Shaul. "A Research Ethics Framework for the Clinical Translation of Healthcare Machine Learning." *American Journal of Bioethics* 22, no. 5 (2022): 8–22.

McCradden, Melissa D., James A. Anderson, and Randi Zlotnik Shaul. "Accountability in the Machine Learning Pipeline: The Critical Role of Research Ethics Oversight." *American Journal of Bioethics* 20, no. 11 (2020): 40–42.

McCradden, Melissa D., Elizabeth A. Stephenson, and James A. Anderson. "Clinical Research Underlies Ethical Integration of Healthcare Artificial Intelligence." *Nature Medicine* 26, no. 9 (2020): 1325–26.

McDougall, Rosalind J. "Computer Knows Best? The Need for Value-Flexibility in Medical AI." *Journal of Medical Ethics* 45, no. 3 (2019): 156–60.

Meier, Lukas J., Alice Hein, Klaus Diepold, and Alena Buyx. "Algorithms for Ethical Decision-Making in the Clinic: A Proof of Concept." *American Journal of Bioethics* 22, no. 7 (2022): 4–20.

Melis, Luca, Congzheng Song, Emiliano De Cristofaro, and Vitaly Shmatikov. "Exploiting Unintended Feature Leakage in Collaborative Learning." Paper presented at the 2019 IEEE Symposium on Security and Privacy (SP), San Franciso, May 23, 2019.

Mittelstadt, Brent Daniel, and Luciano Floridi. "The Ethics of Big Data: Current and Foreseeable Issues in Biomedical Contexts." *Science and Engineering Ethics* 22, no. 2 (2016): 303–41.

Moffett, Peter, and Gregory Moore. "The Standard of Care: Legal History and Definitions; the Bad and Good News." *Western Journal of Emergency Medicine* 12, no. 1 (2011): 109.

Moore, T. J., H. Zhang, G. Anderson, and G. C. Alexander. "Estimated Costs of Pivotal Trials for Novel Therapeutic Agents Approved by the US Food and Drug Administration, 2015–2016." *JAMA Internal Medicine* 178, no. 11 (2018): 1451–57.

Morley, Jessica, Caio C. V. Machado, Christopher Burr, Josh Cowls, Indra Joshi, Mariarosaria Taddeo, and Luciano Floridi. "The Ethics of AI in Health Care: A Mapping Review." *Social Science & Medicine* 260 (2020): 113172.

Morris, Zoë Slote, Steven Wooding, and Jonathan Grant. "The Answer Is 17 Years, What Is the Question: Understanding Time Lags in Translational Research." *Journal of the Royal Society of Medicine* 104, no. 12 (2011): 510–20.

Nasr, Milad, Reza Shokri, and Amir Houmansadr. "Comprehensive Privacy Analysis of Deep Learning: Passive and Active White-Box Inference Attacks against Centralized and Federated Learning." Paper presented at the 2019 IEEE Symposium on Security and Privacy (SP), San Franciso, May 23, 2019.

National Commission for the Protection of Human Subjects of Biomedical and Behavioral Research. *The Belmont Report: Ethical Principle and Guidelines for the Protection of Human Research Subjects.* Office of the Secretary, Department of Health and Human Services.

April 18, 1979. www.hhs.gov/ohrp/regulations-and-policy/belmont-report/read-the-belmont-report/index.html.

"NCA—Intracranial Stenting and Angioplasty (Cag-00085r2)—Decision Memo." Accessed November 22nd, 2022. www.cms.gov/medicare-coverage-database/view/ncacal-decision-memo.aspx?proposed=Y&ncaid=177.

"NJ A1494." Bill Track 50. Accessed September 8, 2023. www.billtrack50.com/BillDetail/1171771#:~:text=Bill%20Summary,consent%20to%20the%20invasive%20examination.

Obermeyer, Ziad, Brian Powers, Christine Vogeli, and Sendhil Mullainathan. "Dissecting Racial Bias in an Algorithm Used to Manage the Health of Populations." *Science* 366, no. 6464 (2019): 447–53.

Onesimu, J. Andrew, J. Karthikeyan, and Yuichi Sei. "An Efficient Clustering-Based Anonymization Scheme for Privacy-Preserving Data Collection in Iot Based Healthcare Services." *Peer-to-Peer Networking and Applications* 14 (2021): 1629–49.

Ozrazgat-Baslanti, T., Y. Ren, R. Islam, H. Hashemighouchani, Matthew M. Ruppert, S. Miao, Tyler J. Loftus, et al. "Development and Validation of a Race-Agnostic Computable Phenotype for Kidney Health in Adult Hospitalized Patients." *Frontiers in Artificial Intelligence.* Forthcoming.

Pandit, Jay A., Jennifer M. Radin, Giorgio Quer, and Eric J Topol. "Smartphone Apps in the COVID-19 Pandemic." *Nature Biotechnology* 40, no. 7 (2022): 1013–22.

Parasuraman, Raja, and Dietrich H. Manzey. "Complacency and Bias in Human Use of Automation: An Attentional Integration." *Human Factors* 52, no. 3 (2010): 381–410.

Parikh, Ravi B., and Lorens A. Helmchen. "Paying for Artificial Intelligence in Medicine." *NPJ Digital Medicine* 5, no. 1 (2022): 63.

Parikh, Ravi B., Stephanie Teeple, and Amol S. Navathe. "Addressing Bias in Artificial Intelligence in Health Care." *JAMA* 322, no. 24 (2019): 2377–78.

Parums, Dinah V. "Artificial Intelligence (AI) in Clinical Medicine and the 2020 CONSORT-AI Study Guidelines." *Medical Science Monitor:*

International Medical Journal of Experimental and Clinical Research 27 (2021): e933675-1.

Pellegrini, Carlos A. "Trust: The Keystone of the Patient-Physician Relationship." *Journal of the American College of Surgeons* 224, no. 2 (2017): 95–102.

Peng, Le, Gaoxiang Luo, Andrew Walker, Zachary Zaiman, Emma K. Jones, Hemant Gupta, Kristopher Kersten, et al. "Evaluation of Federated Learning Variations for COVID-19 Diagnosis Using Chest Radiographs from 42 US and European Hospitals." *Journal of the American Medical Informatics Association* 30, no. 1 (2023): 54–63.

Perez, Caroline Criado. *Invisible Women: Data Bias in a World Designed for Men.* New York: Abrams, 2019.

Pilkington, Bryan, and Charles Binkley. "Disproof of Concept: Resolving Ethical Dilemmas Using Algorithms." *American Journal of Bioethics* 22, no. 7 (2022): 81–83.

Popejoy, Alice B., Deborah I. Ritter, Kristy Crooks, Erin Currey, Stephanie M. Fullerton, Lucia A. Hindorff, Barbara Koenig, et al. "The Clinical Imperative for Inclusivity: Race, Ethnicity, and Ancestry (Rea) in Genomics." *Human Mutation* 39, no. 11 (2018): 1713–20.

Price, W. Nicholson, Sara Gerke, and I. Glenn Cohen, "Potential Liability for Physicians Using Artificial Intelligence," *JAMA* 322, no. 18 (2019): 1765–66.

Prosperi, Mattia, Yi Guo, and Jiang Bian. "Bagged Random Causal Networks for Interventional Queries on Observational Biomedical Datasets." *Journal of Biomedical Informatics* 115 (2021): 103689.

Quill, Timothy E., and Christine K. Cassel. "Nonabandonment: A Central Obligation for Physicians." *Annals of Internal Medicine* 122, no. 5 (1995): 368–74.

Quinonero-Candela, Joaquin, Masashi Sugiyama, Anton Schwaighofer, and Neil D. Lawrence. *Dataset Shift in Machine Learning.* Cambridge, MA: MIT Press, 2008.

Reddy, Sandeep, Wendy Rogers, Ville-Petteri Makinen, Enrico Coiera, Pieta Brown, Markus Wenzel, Eva Weicken, et al. "Evaluation Framework to Guide Implementation of AI Systems into Healthcare Settings." *BMJ Health & Care Informatics* 28, no. 1 (2021): e100444.

Reynolds, Joel Michael, Charles E. Binkley, and Andrew Shuman. "The Complex Relationship between Disability Discrimination and Frailty Scores." *American Journal of Bioethics* 21, no. 11 (2021): 74–76.

Ribeiro, Marco Tulio, Sameer Singh, and Carlos Guestrin. "'Why Should I Trust You?'" Explaining the Predictions of Any Classifier." In *Proceedings of the 22nd ACM SIGKDD International Conference on Knowledge Discovery and Data Mining*, 1135–44. Association for Computing Machinery, 2016. https://www.kdd.org/kdd2016/papers/files/rfp0573-ribeiroA.pdf.

Richardson, Jordan P., Cambray Smith, Susan Curtis, Sara Watson, Xuan Zhu, Barbara Barry, and Richard R. Sharp. "Patient Apprehensions about the Use of Artificial Intelligence in Healthcare." *NPJ Digital Medicine* 4, no. 1 (2021): 140.

Rojas, Juan C., John Fahrenbach, Sonya Makhni, Scott C. Cook, James S. Williams, Craig A. Umscheid, and Marshall H. Chin. "Framework for Integrating Equity into Machine Learning Models: A Case Study." *Chest* 161, no. 6 (2022): 1621–27.

Rojas, Juan C., Mario Teran, and Craig A. Umscheid. "Clinician Trust in Artificial Intelligence: What Is Known and How Trust Can Be Facilitated." *Critical Care Clinics* 39, no. 4 (2023): 769–82.

Rosenfeld, Amir, Richard Zemel, and John K Tsotsos. "The Elephant in the Room." arXiv, August 9, 2018. https://doi.org/10.48550.arXiv.1808.03305.

Russell, Regina G., Laurie Lovett Novak, Mehool Patel, Kim V. Garvey, Kelly Jean Thomas Craig, Gretchen P. Jackson, Don Moore, et al. "Competencies for the Use of Artificial Intelligence–Based Tools by Health Care Professionals." *Academic Medicine* 98, no. 3 (2023): 348–56.

Shahian, David M., Jeffrey P. Jacobs, Vinay Badhwar, Paul A. Kurlansky, Anthony P. Furnary, Joseph C. Cleveland Jr., Kevin W. Lobdell, et al. "The Society of Thoracic Surgeons 2018 Adult Cardiac Surgery Risk Models: Part 1—Background, Design Considerations, and Model Development." *Annals of Thoracic Surgery* 105, no. 5 (2018): 1411–18.

Shimabukuro, David W., Christopher W. Barton, Mitchell D. Feldman, Samson J. Mataraso, and Ritankar Das. "Effect of a Machine

Learning-Based Severe Sepsis Prediction Algorithm on Patient Survival and Hospital Length of Stay: A Randomised Clinical Trial." *BMJ Open Respiratory Research* 4, no. 1 (2017): e000234.

Solomon, Neil A., Henry A. Glick, Christopher J. Russo, Jason Lee, and Kevin A. Schulman. "Patient Preferences for Stroke Outcomes." *Stroke* 25, no. 9 (1994): 1721–25.

Springer, Aaron, and Steve Whittaker. "Making Transparency Clear: The Dual Importance of Explainability and Auditability." In *Joint Proceedings of the ACM IUI 2019 Workshop, Los Angeles, CA, USA, March 20, 2019.* Association for Computing Machinery, 2019. https://ceur-ws.org/Vol-2327/IUI19WS-IUIATEC-5.pdf.

Steinhubl, Steven R., Dana L. Wolff-Hughes, Wendy Nilsen, Erin Iturriaga, and Robert M Califf. "Digital Clinical Trials: Creating a Vision for the Future." *NPJ Digital Medicine* 2, no. 1 (2019): 126.

Stern, A. D., A. Goldfarb, T. Minssen, and W. N. Price II. AI Insurance: How Liability Insurance Can Drive the Responsible Adoption of Artificial Intelligence in Health Care. *NEJM Catalyst Innovations in Care Delivery* 3, no. 4 (2022): CAT-21.

Subbaswamy, Adarsh, and Suchi Saria. "From Development to Deployment: Dataset Shift, Causality, and Shift-Stable Models in Health AI." *Biostatistics* 21, no. 2 (2020): 345–52.

"Surgical Risk Calculator." American College of Surgeons. Accessed October 9, 2023. https://riskcalculator.facs.org/RiskCalculator/PatientInfo.jsp.

Topol, Eric J. "Welcoming New Guidelines for AI Clinical Research." *Nature Medicine* 26, no. 9 (2020): 1318–20.

U.S. Food and Drug Administration. "Guidance for Industry: E9 Statistical Principles for Clinical Trials." Accessed November 22, 2022. www.fda.gov/regulatory-information/search-fda-guidance-documents/e9-statistical-principles-clinical-trials.

van der Ven, Ward H., Denise P. Veelo, Marije Wijnberge, Björn J. P. van der Ster, Alexander P. J. Vlaar, and Bart F. Geerts. "One of the First Validations of an Artificial Intelligence Algorithm for Clinical Use: The Impact on Intraoperative Hypotension Prediction and Clinical Decision-Making." *Surgery* 169, no. 6 (22021): 1300–1303.

https://doi.org/https://doi.org/10.1016/j.surg.2020.09.041. https://www.sciencedirect.com/science/article/pii/S0039606020307728.

Varkey, Basil. "Principles of Clinical Ethics and Their Application to Practice." *Medical Principles and Practice* 30, no. 1 (2021): 17–28.

Vollmer, Sebastian, Bilal A. Mateen, Gergo Bohner, Franz J. Király, Rayid Ghani, Pall Jonsson, Sarah Cumbers, et al. "Machine Learning and Artificial Intelligence Research for Patient Benefit: 20 Critical Questions on Transparency, Replicability, Ethics, and Effectiveness." *BMJ* 368 (2020): l6927.

Vyas, Darshali A., Leo G. Eisenstein, and David S. Jones. "Hidden in Plain Sight—Reconsidering the Use of Race Correction in Clinical Algorithms." *Massachusetts Medical Society* 383, no. 9 (2020): 874–82.

Wagner, Richard F., Jr., Abel Torres, and Steven Proper. "Informed Consent and Informed Refusal." *Dermatologic Surgery* 21, no. 6 (1995): 555–59.

Wakabayashi, Daisuke. "Google and the University of Chicago Are Sued over Data Sharing." *New York Times*, June 26, 2019.

Wang, Zhibo, Mengkai Song, Zhifei Zhang, Yang Song, Qian Wang, and Hairong Qi. "Beyond Inferring Class Representatives: User-Level Privacy Leakage from Federated Learning." Paper presented at the IEEE INFOCOM 2019–IEEE Conference on Computer Communications, Paris, April 29–May 2, 2019.

Wei, Wenqi, Ling Liu, Margaret Loper, Ka-Ho Chow, Mehmet Emre Gursoy, Stacey Truex, and Yanzhao Wu. "A Framework for Evaluating Gradient Leakage Attacks in Federated Learning." arXiv, April 23, 2020. https://doi.org/10.48550/arXiv.2004.10397.

Wetsman, Nicole. "Hospitals Are Selling Treasure Troves of medical Data—What Could Go Wrong." The Verge, June 23, 2021. https://www.theverge.com/2021/6/23/22547397/medical-records-health-data-hospitals-research.

Wiens, Jenna, Suchi Saria, Mark Sendak, Marzyeh Ghassemi, Vincent X. Liu, Finale Doshi-Velez, Kenneth Jung, et al. "Do No Harm: A Roadmap for Responsible Machine Learning for Health Care." *Nature Medicine* 25, no. 9 (2019): 1337–40.

Wijnberge, Marije, Bart F. Geerts, Liselotte Hol, Nikki Lemmers, Marijn P. Mulder, Patrick Berge, Jimmy Schenk, et al. "Effect of a Machine Learning–Derived Early Warning System for Intraoperative Hypotension vs. Standard Care on Depth and Duration of Intraoperative Hypotension During Elective Noncardiac Surgery: The Hype Randomized Clinical Trial." *JAMA* 323, no. 11 (2020): 1052–60. https://doi.org/10.1001/jama.2020.0592.

World Health Organization. "Ethics and Governance of Artificial Intelligence for Health." *WHO Guidance*, June 28, 2021. www.who.int/publications/i/item/9789240029200.

Wratschko, Katharina. *Strategic Orientation and Alliance Portfolio Configuration.* Wiesbaden: Springer, 2009.

Zawati, Ma'N., and Michael Lang. "What's in the Box? Uncertain Accountability of Machine Learning Applications in Healthcare." *American Journal of Bioethics* 20, no. 11 (2020): 37–40.

Zicari, Roberto V., John Brodersen, James Brusseau, Boris Düdder, Timo Eichhorn, Todor Ivanov, Georgios Kararigas, et al. "Z-Inspection®: A Process to Assess Trustworthy AI." *IEEE Transactions on Technology and Society* 2, no. 2 (2021): 83–97.

Zivanovic, N. Nina. "Medical Information as a Hot Commodity: The Need for Stronger Protection of Patient Health Information." *Intellectual Property Law Bulletin* 19 (2014): 183.

INDEX

Founded in 1893,
UNIVERSITY OF CALIFORNIA PRESS
publishes bold, progressive books and journals on topics in the arts, humanities, social sciences, and natural sciences—with a focus on social justice issues—that inspire thought and action among readers worldwide.

The UC PRESS FOUNDATION
raises funds to uphold the press's vital role as an independent, nonprofit publisher, and receives philanthropic support from a wide range of individuals and institutions—and from committed readers like you. To learn more, visit ucpress.edu/supportus.